THE LIE of THE LAND

POWERHOUSE

NATIONAL CENTRE FOR AUSTRALIAN STUDIES

MONASH UNIVERSITY

IDEAS FOR AUSTRALIA

THE LIE OF THE LAND

PUBLISHER
'Ideas for Australia Program 1991–92' and the National Centre for Australian Studies, (Clayton Campus) Monash University, 3168 AUSTRALIA
in association with the Powerhouse Museum, Sydney

EDITORS
Annette Shiell & Ann Stephen

PRODUCTION EDITOR
Jane Yule

PRODUCTION ASSISTANT
Cathy Edmonds

DESIGN
Cozzolino/Ellett Design D'Vision, Hawthorn East, Vic.
Tel: (03) 882 9711 Fax: (03) 882 9779

IMAGE SETTING
Bandaid Productions, Fitzroy, Vic.
Tel: (03) 419 5385 Fax: (03) 419 4106

PRINTING
Photo-Offset Productions, Fitzroy, Vic.
Tel: (03) 419 5344 Fax: (03) 417 6521

First Published 1992

ISBN 0 7326 0298 X

EXHIBITION DATES

Powerhouse Museum, Sydney, NSW
21 February 1992 – 20 July 1992

National Library of Australia, Canberra, ACT
17 August 1992 – 25 October 1992

National Wool Museum, Geelong, Victoria
21 November 1992 – 21 February 1993

CO-CURATORS
Annette Shiell and Ann Stephen

CONTENTS

ACKNOWLEDGMENTS

Financial assistance from the 'Ideas for Australia Program 1991–1992' has made this exhibition possible. The work of the Powerhouse Museum in mounting this exhibition is gratefully acknowledged, particularly the labours of Michelle Andringa, Janine Beardmore, Julie Donaldson, Susan McMunn, Jai Paterson, Kathleen Phillips, Colin Rowan, Jennifer Sanders, Anne Slam, Karin Vesk, Kimberley Webber, and especially the curatorial work of Ann Stephen.

The National Centre for Australian Studies also wishes to acknowledge the work of the exhibition designer Jisuk Han, and the catalogue designer Andrew Hook, as well as the generosity and cooperation of the following institutions and individuals who have assisted with the exhibition: Artarmon Galleries and Phillip Brackenreg; Art Gallery of New South Wales and Amanda Beresford; Associated Pulp and Paper Mills and Lyndon Humphrey; Australian Airlines and Glynn Cumming; Australian Conservation Foundation and Felicity Wishart; Australian Posters and Mike Humphries; The Broken Hill Proprietary Company Limited and Stedman Ellis; Chiat/Day/Mojo and Gawen Rudder, Lindsay Lewis and David Steindl; Coca-Cola Pty Ltd and Bevan Williams; Garage Graphix; Green Ant and Chips Mackinolty; McCann Erickson; Mambo and Wayne Golding and Dare Jennings; Mental as Anything; The National Library of Australia and Harry McCarthy; Redback Graphix; Red Planet and Paul Macgregor; Rimfire Films; State Library of Victoria and Linda Notley; Streets Ice-cream Pty Ltd; Tasmaniana Collection, State Library of Tasmania and Tony Marshall; University of Sydney Union Art Collection and Nick Vickers; Victorian Association of Forest Industries and Graeme Gooding; The Wilderness Society (Victorian Branch) and Michael Fogarty; Geoffrey Batchen; Cameron Britt; Mimmo Cozzolino; Ken and Judy Done; Phil Dwyer; Jack Frawley; Moya McFadzean; Marie McMahon; Reg Mombassa; Tracey Naughton; Nicholas Ruskin; Jacky Talbot; Hertz Walpole; Annette Welkamp; Paul Worstead. The Australian Culture Workshop group held a one-day display entitled 'The Lie of the Land' in conjunction with a seminar at the State Library of NSW in March 1990.

Essay contributions to the catalogue are gratefully acknowledged as is the preparation work by John Arnold, Cathy Edmonds, Katie Purvis, Carol Somogyi, Jane Yule, and unless otherwise credited, the photography by Susan Stafford from the Powerhouse Museum. Thanks also to the organisations and individuals who generously granted permission for the reproduction of works in the catalogue.

Annette Shiell
National Centre for Australian Studies, Monash University

FOREWORD

This exhibition represents a widely shared interest in the ways that Australian landscapes have been presented and massaged for mass consumption. In its current form it owes its origins to a sighting of an advertisement for an alcoholic beverage, Malibu, displayed on Sydney suburban railway station hoardings in 1990. The advertisement showed Uluru surrounded by a great inland sea, and was accompanied by the slogan: 'Malibu, Wherever You Are It's a Tropical State' (Exhibit no. 2).

In the 1980s landscape imagery became one of the most dominant ways to sell ideas, products and services, and many groups in our society adapted landscape themes for their own purposes. These range from the propagation of spiritual values to political statements, from the promotion of new products to concerns about how best to benefit from and protect our environment. The range of purpose is paralleled by the range of mediums which convey landscape imagery. To give this imagery historical context the curators have selected four paintings and one lithograph for the five landscape areas they concentrate on, namely: coastal; suburban; pastoral; rainforest; and the outback.

The exhibition itself draws on an enormous variety of mediums, from the mainstream commercial world of billboard, press and television advertising to less widely distributed, but often equally influential, imagery produced by artists, designers and community groups using the poster as their most common means of expression. Even humble tea-towels and souvenirs convey pervasive landscape themes, reflecting and reinforcing popular attitudes about Australia.

When the 'Ideas for Australia' committee, chaired by Donald Horne, agreed to support an exhibition on ways of interpreting and displaying the contemporary Australian landscape, we approached the Powerhouse Museum to see if it would be interested in being involved in the development of such an exhibition. **The Lie of the Land** is the result of that collaboration, and is jointly curated by Ann Stephen from the Powerhouse Museum and Annette Shiell from the National Centre for Australian Studies.

The 'Ideas for Australia' committee hopes the exhibition will stimulate debate about the use of landscape imagery in political and commercial advertising, and encourage discussion about the formulation of local, regional and national stereotypes. The committee also hopes it will be of particular interest to high school students and young adults, the principal target audience for many of these images.

Peter Spearritt
National Centre for Australian Studies, Monash University

The land is the centre of Australian consciousness. Australia first emerged in European consciousness as the world's only island continent, starting its imaginary and then cartographic life as the great south land and becoming, by the time of European exploration, *Terra Australis Incognita.* The world's only island continent has also been, at least for the last 50,000 years, the least populated continent. This was the case in 1492, in 1788 and in 1992, 500 years after Christopher Columbus.

The importance of the land in conceptions of Australia, by both Aboriginal people and Australians of European and Asian origins, stems in part from questions of size and the sparsity of settlement. Most Australians now realise, even if they are uncomfortable with the knowledge, that this is a highly urbanised society, with more than 80 per cent of the population living in six coastal capitals and their adjacent satellite settlements. Only one inland city, Canberra, has a population of more than 100,000. Almost 14 out of 17 million people live in less than 100,000 square kilometres of a land mass of over 7.5 million square kilometres. Nine-tenths of the continent remains almost uninhabited, much as it was in 1788.

The perception of Australia as a virtually uninhabited continent, with unforeclosed land as its greatest asset, captured the European imagination. Europeans, who for centuries had been developing systems of land ownership and leasing, quickly moved to impose such a system in Australia, in much the same way as they restructured land ownership in the United States, Africa and South America. Most of the European settlers had little understanding of how Aboriginal people used the land, nor did they understand their notion of territoriality. The phrase *Terra Australis Nullius* summed up the notion that Australia was an untouched continent and could be carved up. A handful of early Europeans did observe and remark on Aboriginal notions of territoriality. In 1850 Bishop Rosendo Salvado of the Spanish mission at New Norcia, north of Perth, wrote:

Every individual has his own territory for hunting, gathering gum and picking up yams, and the rights he has here are respected as sacred... Consequently each family regards one particular district as belonging exclusively to itself, though the use of it is freely shared by nearby families.[1]

Australia, in terms of European law, started out with one owner, the British Government, which chose not to sign a treaty with any particular tribes nor to acknowledge Aboriginal peoples' claims on the land. In the interests of encouraging settlement and consolidating its hold on the continent, the governments of each colony promptly gave, sold or leased vast tracts of land to immigrant farmers, land speculators, pastoralists, and the churches.[2] The colonial governments, whether as British satellites or later as self-governing entities, also kept a lot for themselves, either for defence

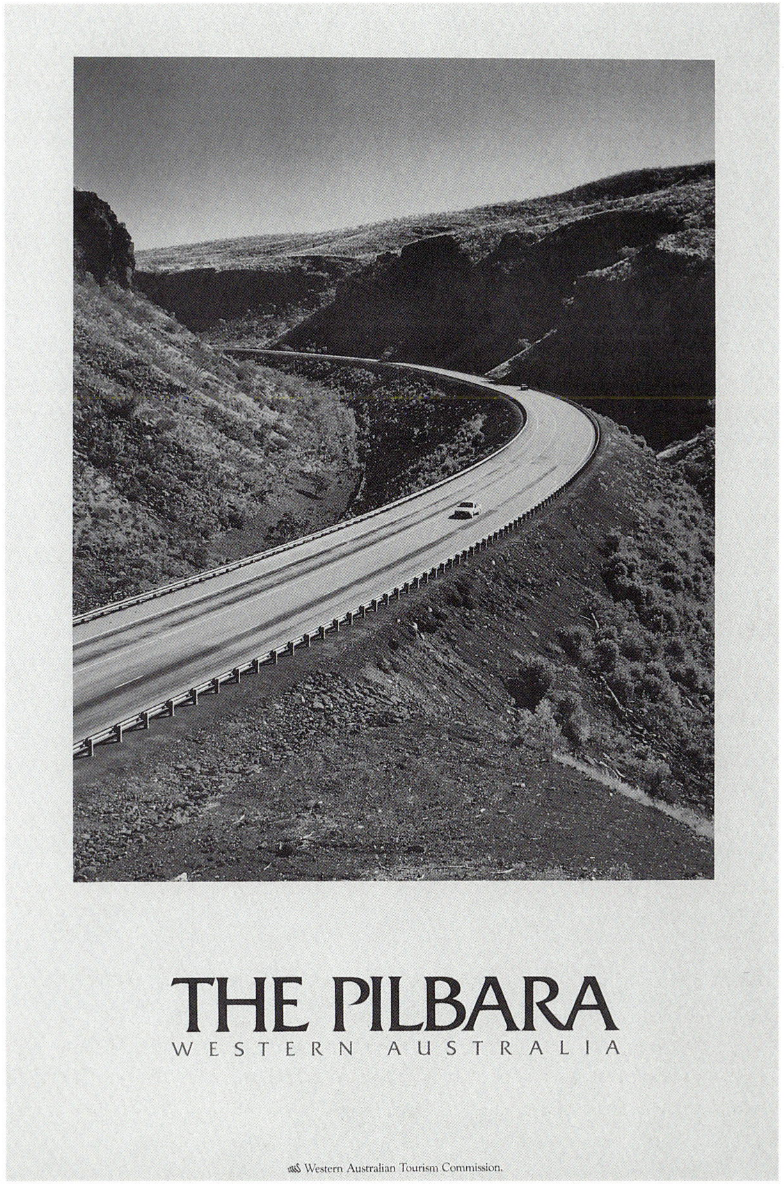

***The Pilbara*, c. 1990 Juxtaposes barren landscape with the might of modern transport Copyright: Western Australian Tourism Commission Exhibit no: 57**

purposes or because the land was inaccessible or thought to have no economic use. This explains why much of the Australian coastline, many of the mountains and most of the desert remain in public hands.

In a country where neither land nor timber was seen as a scarce resource, land settlement and land clearing proceeded at a startling and ruthless pace. Rainforests were routed for their valuable timbers, eucalypt forests were cut down for housing, firewood and railway building, while tracks, roads and railways linked emerging townships. From the 1860s ringbarking became common among pastoralists looking for a cheap and simple way to increase their grazing lands.[3] The continent had to be conquered. The first agents for overcoming the 'tyranny of distance' were ships and horse-drawn vehicles, which were later supplanted for movement within the continent by railways and trucks.[4] Most marketing images of the Australian bush and the outback continue to juxtapose uninhabited stretches of landscape with the might of modern transport and technology, whether of a majestic highway cutting through the Pilbara or a coal dredger at work in an open cut mine.

OWN THE LAND YOU LOVE

In the twentieth century the hopes of the early European settlers were realised; most Australian families became landowners, even if their only holding was a suburban block of land. In the 1920s the real estate industry pushed slogans such as 'Own a Little of the Land You Love' and by the end of that decade over half of all households did, even if they shared ownership with a bank or building society. Land ownership in Australia peaked in 1971 with over 70 per cent of households owning or buying their residence. Looked at in terms of the lifespan of the average Australian household, the great majority of households will be landowners for some portion of their lives. Land, and the housing that sits on it, is the principal asset of most Australian households and the one most likely to be passed on to succeeding generations.

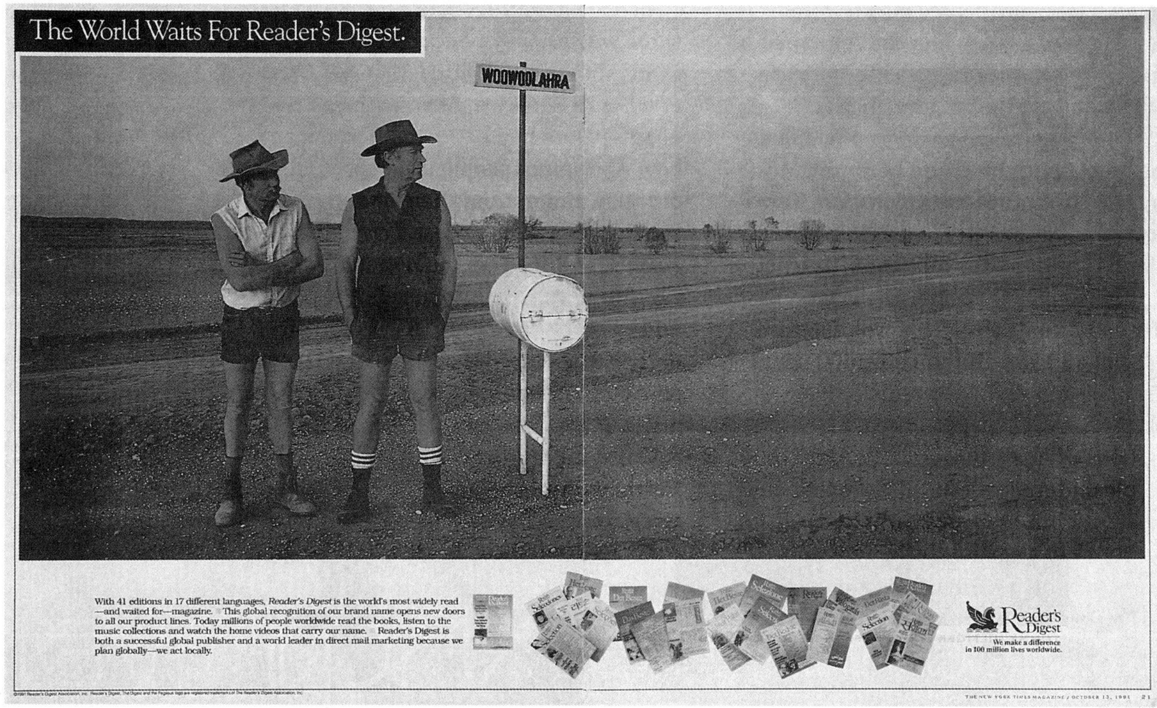

Reader's Digest Advert **Two bushies suggest to Americans that Reader's Digest even penetrates the imaginary outback of Woowoolahra *The New York Times Magazine*, 13 October 1991 Copyright: The Reader's Digest Association, Inc.**

In a society where the single most important characteristic of consumers is that almost three-quarters of the population are outright owners or mortgagees of their homes, the predominance of house and yard in contemporary Australian image-making is not surprising. Political parties use it in campaigns, whether as scare tactics (a capital gains tax on the family home) or as inducements (first home owner schemes).

Advertisements for new housing estates continue to emphasise the dignity of the subdivision pattern and the proximity of the estate to major urban centres. These centres are usually denoted by a symbol of some kind (the Harbour Bridge in Sydney, high-rise buildings elsewhere) even if the city centre is 50 kilometres away. In dominant real estate imagery Australian cities provide the best of both worlds, a backyard for children to play in, and access to the jobs and entertainment provided by urban centres.

The most evocative image-making about suburban Australia has come from photographers, painters and graphic artists, the latter often working in a way which mixes social comment with satire. In their *Older than the Hills*, Eliza Campbell and Judith Lodwick place housewife, washing and Hills Hoist in the foreground, with the hint of triple fronted houses in the middle ground, while ancient Uluru and the outback are at once the background and the landscape setting (Exhibit no. 81a/b). In this vision of the continent every Australian backyard reaches to the Centre.

CONSUMING LANDSCAPES

Most Australian households occupy flat suburban blocks with the majority of the space out the back, but wealthier households and those in smaller cities often manage to grab a bit of topography and even a horizon. The most notable example of this is Sydney Harbour, the site of Australia's most expensive residential real estate. But cities like Launceston and Hobart also offer topographic variation and great views. At the other landscape extreme, some of the poorest and the cheapest housing in Broken Hill — galvanised iron dwellings on the edge of the town — offer the desert beyond, although with less fanfare than the Yulara resort hotels near Uluru.

On retirement, many householders choose to leave their suburban setting and enjoy more attractive, less polluted landscape settings, whether in the mountains (the Dandenongs, the Blue Mountains) or on the coast. Retirees, and those households able to afford a holiday house, often look for a mix of natural terrain and a managed landscape,[5] managed to the extent that roads, golf courses, shops, medical facilities, and the ubiquitous club are within easy reach. In some new tourist developments all these facilities are provided within a 'gated community', where both residents and visitors have to pass through a security check to get into the compound.

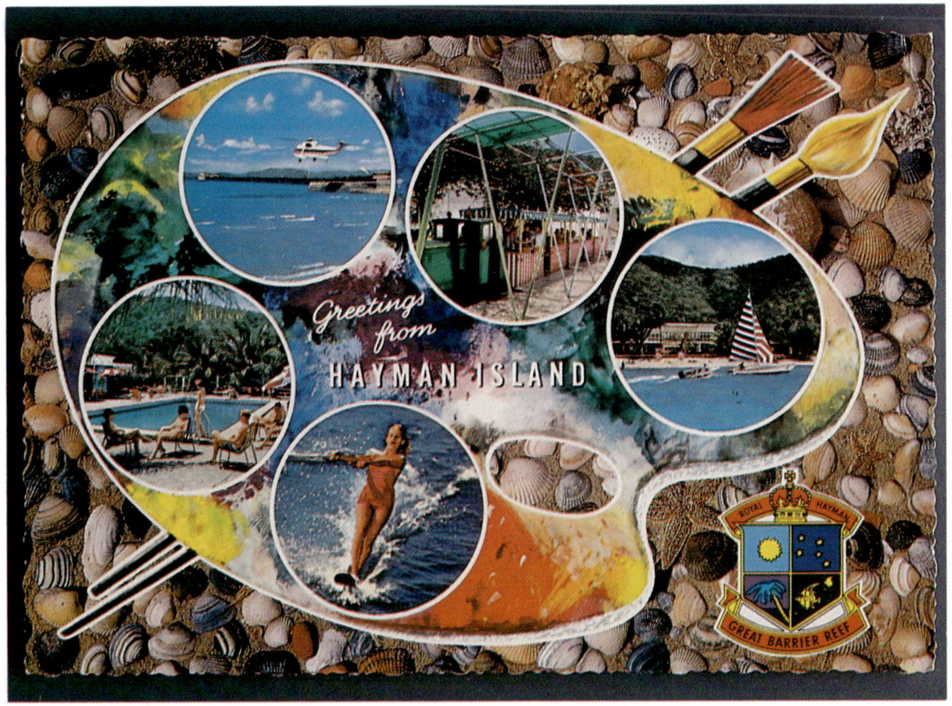

Greetings from Hayman Island, c.1960s postcard

The outline of the Australian coast remains a pervasive image of the nation, both here and in the rest of the world. The coast provided the first sites for European settlement in Australia. Two main images of the Australian coastline now dominate our televisions and print media. The one, of deserted beaches and bushy headlands, is often juxtaposed with the other, of glamorous coastal buildings, good food, beautiful bodies, sexual abundance, and a society in which holidays have replaced work. Coastal landscape is a necessary setting for such images. Generations of Australians have been both subject and object of advertising campaigns for suntan oils, campaigns which have made it clear that the beach is our most populated holiday site.[6]

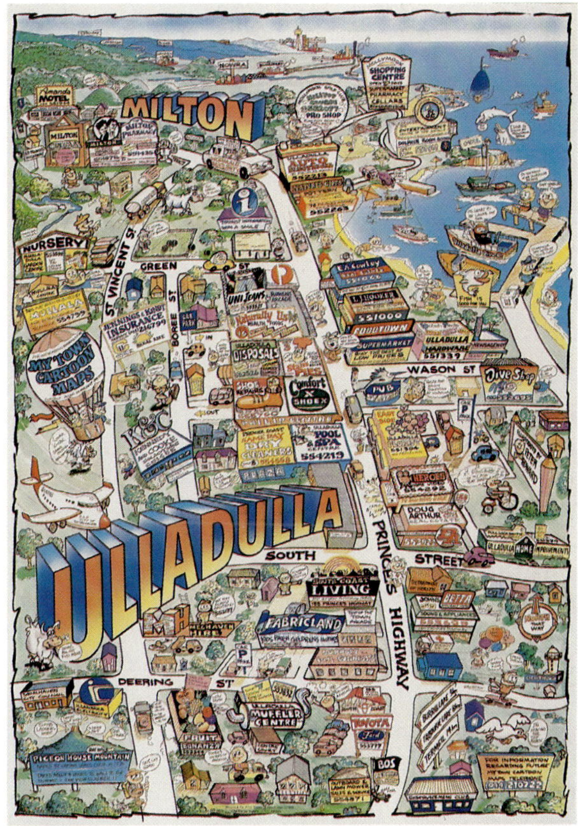

Ulladulla, 1990 Designed by Allan Droscher Real estate agents' view of landscape nirvana Copyright: My Town Cartoon Maps Exhibit no: 26

In the last 30 years the Australian coastline has come under enormous redevelopment pressures. The fishing industry and latterly the leisure marina lobby have demanded safe havens for their vessels, so retaining walls have been built in scores of bays around the nation. But the most notable and the most often depicted development has been of resort accommodation with its attendant shops, cafes, fun parlours, and hair salons. These developments are a direct descendant of the seaside resorts of Victorian and Edwardian England. The difference in Australia is that the landscape impact has been most pronounced since 1960, when high-rise flats first became common on the coast.

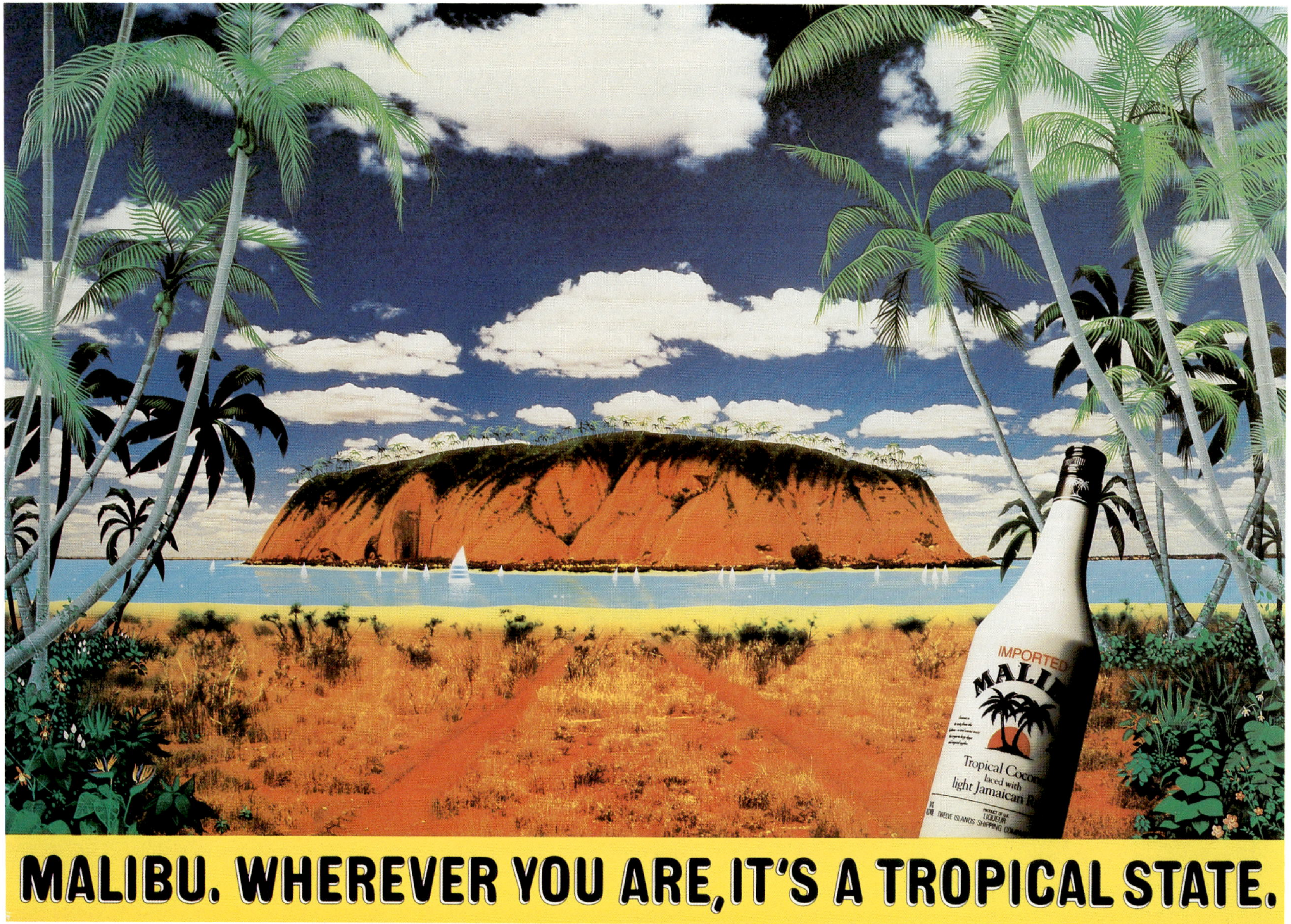

Malibu, Wherever You Are, It's a Tropical State, 1990 **Designed by Clemenger** **Copyright: Swifton Moore** **Exhibit no: 2**

In the 1960s and 1970s these developments, modelled on Waikiki, Miami and the French Riviera, were hailed without artifice. Progress was good. In the marketing climate of the 1980s, when both domestic and overseas consumers were having their 'green' consciousness raised, natural bushland backdrops became more common. The 1991 Hamilton Island advertising campaign demonstrates this perfectly, its high-rise apartment blocks nestled into or dominating (depending on your point of view) the tropical landscape. North of Cairns, low-rise hotels and apartments of the 1980s, inspired by late nineteenth-century colonial architecture, really do blend with their landscape. But even here, developers intent on creating self-contained resorts are still carving out bits of the coastal strip for live-in golf course playgrounds.

Some coastal communities remain so unabashedly pro-development that in depicting themselves they produce landscape images where boom-time building takes precedence over all other landscape features. Confronted by a poster of the NSW south coast towns of Milton and Ulladulla, still in circulation in 1992, one is unlikely to deduce that the settlement boasts unspoiled beaches and the superb hinterland of the Pigeon House mountain. Images like the Ulladulla poster-map unambiguously celebrate the coastal real estate agents' view of nirvana — a landscape made for the car, in which everything can be sold and resold, every site a new commercial opportunity. In the current recession

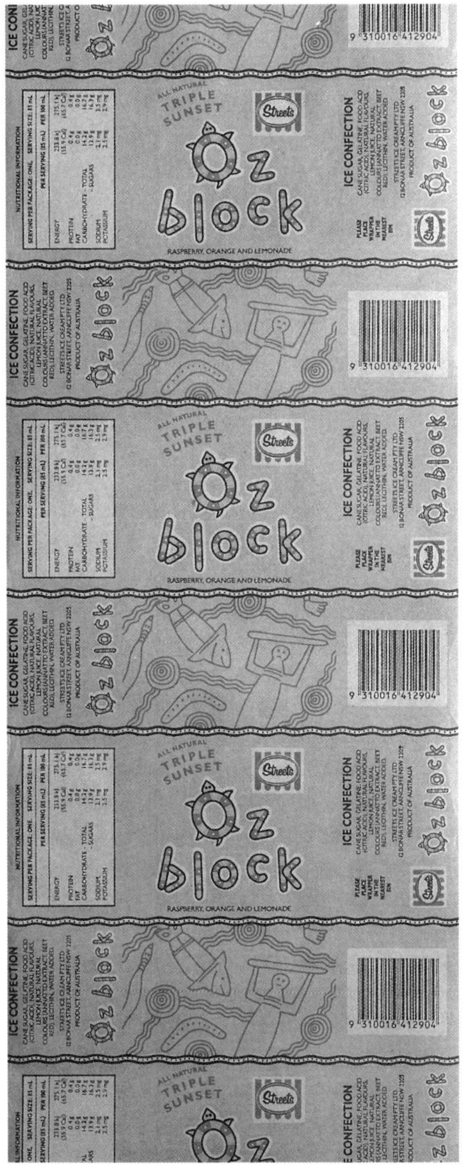

***Oz Block Wrapper*, 1991 Aboriginal imagery is frequently appropriated for new products Copyright: Streets Ice Cream Pty Ltd**

many of this town's shops and flats, like those of other coastal resorts around Australia, are vacant.

Much of the landscape imagery of radical artists in the 1980s, artists who choose to work in the medium of the poster or even the tea-towel, is explicitly anti-commercial in its focus, often taking as one of its themes a satirical attack on an advertising slogan. Campbell and Lodwick's *Come Back to Catholicism* tea-towel, with mounted pontiff, boomerang in hand, and a backdrop of the Australian bush, is a powerful re-working of religious, Aboriginal and landscape imagery. Some may think it in poor taste, but others believe its anti-commercial tone gives it a special marketing appeal.

In the 1980s, landscape imagery that attempted to appropriate and/or acknowledge the place of Aboriginal culture in Australian life came to dominate many advertising campaigns and forms of product identity. In its 'Oz Block' ice confection wrapper, graphic designers for Streets Ice Cream, now a predominantly foreign-owned company, draw on a long tradition of appropriating Aboriginal imagery. Aimed at youthful consumers, these are recognisable images of fish, boomerangs and other Aboriginal designs. Even its typography draws on such imagery. It may be argued that the 'all natural Oz Block' is as much a belittling as a celebration of Aboriginal culture, but it is a long way from 'Mine Tinkit They Fit', a 1940s Pelaco shirt advertising campaign featuring a line drawing of Mulga Fred, a Koori from near Geelong.[7]

THE ADVENT OF COLOUR

With the introduction of colour television in 1975 and the cost reductions and technical improvements in colour magazine printing, advertisers have been able to generate arresting new imagery. During this period advertisements, for even the most practical of goods and services like cars, home loans, and breakfast cereals, have often been shown within a landscape context. This is as true of Australian companies, hoping to cash in on nationalist sympathies, as it is of foreign companies wishing to market their commitment to Australia. The landscape usually provides the background to the advertisement, whether it is a rural road, a leafy suburban backyard or a spectacular bit of coastline; but frequently the landscape also conveys the main message, typically about lifestyle. Particular landscapes can be used to convey a sense of good, natural, even classy lives. CSR's 'Sugar — a Natural Part of Life' campaign emphasises the growing of sugar cane under sunny skies and celebrates tender moments in the extended family.

Two of the most telling uses of colour in recent years have been the National South-West Coalition's advertisement 'Could You Vote for a Party that Will Destroy This?' (see pages 46-48 of this catalogue), and the Aboriginal Flag (see pages 30-31). The Franklin River advertisement was the first full colour political advertisement to appear in an Australian daily newspaper. It used its rainforest/wilderness theme to spectacular effect, signalling that environmental issues had become a major arena of political disputation. The potency of such images has been confirmed by the Forest Industries' television, press and billboard campaigns which frequently rely on images of rainforest re-growth.

Like many of the influential commercial and political images of the 1980s the Franklin Dam advertisement was a very carefully crafted photograph. Improvements in printing technology in the 1960s and 1970s led to photographic images displacing graphic images in most areas of advertising. Weekly and monthly periodicals abandoned their graphic covers for photographs, and commercial artists found themselves restricted to art direction, layout and typographical decisions, as distinct from their former job of creating telling, central images. Realistic and naturalistic images became the dominant way to portray the Australian landscape.[8] Overseas tourists and Australians proud of their land were bombarded with coffee table books of colour photographs showing the diversity of the land and its people. But the net effect was to produce a very limited number of landscape stereotypes: high-rise buildings in the capital cities, suburbanites at the beach, stockmen in the outback, miners at the coalface and motley crowds at the football or the cricket.[9]

In the 1980s graphic designers/artists have made a dramatic comeback, as evidenced by the influence and popularity of images created by Ken Done and Martin Sharp among others. At the same time commercial photographers have been prepared to tackle more abstract themes, including a great deal of experimentation with landscape imagery. In the world of travel posters, landscape imagery varies from mug shots of Gold Coast high-rise buildings with their coastal foreground and rainforest hinterland, to soft focus close ups of 'wilderness areas', whether a Gippsland beach or a rocky outcrop in the Northern Territory.[10]

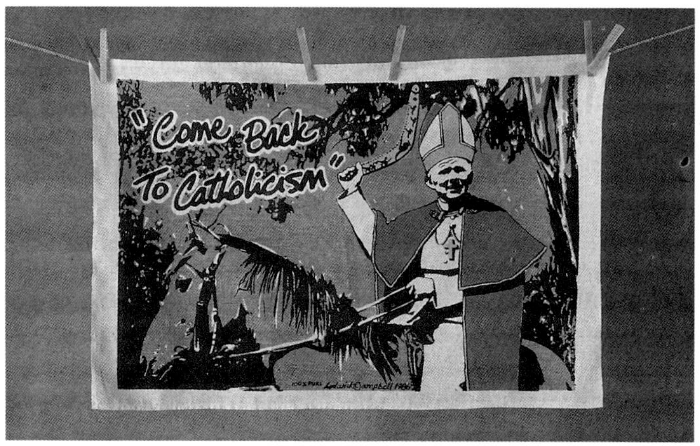

Come Back to Catholicism, 1986 Designed by Eliza Campbell & Judith Lodwick A powerful re-working of religious, Aboriginal & landscape imagery Exhibit no: 82

In producing landscape images, the photographers and graphic designers of the 1980s draw on a rich heritage of attempts to represent Australian landscapes. Their influences are diverse, from the wide range of Aboriginal painting to the work of Margaret Preston, Fred Williams and Brett Whiteley. Some advertising executives, in search of more romantic representations, have delved back into the Heidelberg School. Thus we have the famous Kit-Kat chocolate television advertisement based around one of Frederick McCubbin's nostalgic bushland paintings.

Of all the symbols that a nation presents to the world, few have more permanance than the official flag. Although the Australian flag is not as ubiqituous in our official buildings and classrooms as the Stars and Stripes is in America, it nonetheless has wide distribution, especially during time of war. Most national flags draw on heraldic symbolism (the Union Jack symbolises the union of three ancient kingdoms), federalism (the United States), flora (the Canadian maple leaf), political symbols (the hammer and sickle of the former USSR), blocks of colour (the Italian and the French), or celestial symbols (Japan, Pakistan and Laos).

The Australian flag is the result of a competition among professional and amateur graphic artists held by the Federal Government at the time of federation. The official Australian Flag incorporates the imperial Union Jack with a European's view of what is distinctive about Australia, namely that the Southern Cross can be seen here and not in the northern hemisphere.[11] This flag was submitted to King Edward for approval in 1902. The Aboriginal Flag, on the other hand, attempts to symbolise both land and people. It relies on simplicity of design, boldness of colour and directness of theme — black for the Aboriginal people, yellow for the sun and red for the earth — to achieve its impact. It has much to say about the lie of the land.

ENDNOTES

[1] Quoted in Sylvia Hallam 'Peopled Landscapes: Original Australian Land Holding and Land Management in South-western Australia', *Westerly*, No.4, p. 90, December 1991

[2] See D. N. Jeans 'The Incorporation of Australia', in D. N. Jeans (ed) *Space and Society*, Sydney University Press, 1987

[3] G. Bolton *Spoils and Spoilers*, pp. 42-44, Allen & Unwin, Sydney, 1981

[4] See G. Blainey *The Tyranny of Distance*, Sun Books, Melbourne, 1970

[5] Cf Tony Hiss *The Experience of Place*, p. 140, Random House, New York, 1991

[6] See G. Dutton *Sun, Sea, Surf and Sand — The Myth of The Beach*, Oxford University Press, Melbourne, 1985

[7] Mimmo Cozzolino *Symbols of Australia*, p.129, Penguin, Ringwood, 1980

[8] A useful record of trends in commercial art and photography are the annuals produced by the Australian Commercial and Industrial Artists Association (ACIAA) in the 1960s and *The Australian Advertising Folio*, North Sydney, 1984

[9] See, for instance, the photographs in Robert B. Goodman and George Johnston *The Australians*, Rigby, Adelaide, 1966 & *This is Australia*, a best selling Ure Smith coffee table extravaganza, Sydney, 1975 and subsequent reprintings.

[10] See P. Spearritt 'Sites and Sights: Australian Travel Posters 1909—1990' in *Trading Places* (exhibition catalogue), pp. 3-12, Monash University Gallery, 1990

[11] The Southern Cross had already been incorporated in the Victorian colonial badge, in the blue silk banner hoisted by miners at the Eureka Stockade and in the white ensign of the federation movement. See Gavin Souter *Lion and Kangaroo*, p.190, Fontana, Sydney, 1978

FROM WILDERNESS TO NATURE STRIP — Ann Stephen

This exhibition, **The Lie of the Land**, considers how the Australian landscape is portrayed by contemporary mass media and popular culture. It offers an opportunity to see how such images enliven our sense of place and locality, time, and history. As the title implies there is no such thing as a 'natural' landscape. Indeed, many non-Western societies do not make the 'culture/nature' distinction which is inherent in much of our thinking. The word 'landscape' is of Dutch origin meaning to shape the land, and it is this sense of construction and manufacture of imagery that the exhibition addresses.

The Lie of the Land focuses on the different ways land in Australia is represented; our relationship to it; and the ways in which sites for wilderness, tourism, housing, industry, and leisure are depicted. Such diverse ways of seeing the same landscape or site gives us a different perspective towards it and forces us to value what we see in quite different ways.

The exhibition counterposes imagery that usually exists in separate spheres. Posters from various community movements which seek to change the physical and social landscape have been juxtaposed with advertisements plucked from the market place. Exhibits are organised around five popular landscape sites: coastal, suburban, pastoral, rainforest, and the outback. These represent places where geographical concepts intersect with Western constructs of social space and time.

COAST — PARADISE LOST

'He's my blond-headed stompie-wompie
Real gone surfer boy'
Little Pattie, 1963

Our shoreline is unique in that it marks out Australia as both island and nation. We send more postcards of beach and coastal views, partly because most of us holiday there. Often, however, the photograph is taken from the point of view of someone looking out to the blue beyond with their backs to the land, thereby giving a stronger sense of the conventions of the imagery than the geography of the coastline.

The coast is the most 'sexed' of landscapes, with nubile bodies never far from view. Max Dupain's *Sunbather,* whose sweaty sensuality has become an icon of this genre, suggests how the body has come to represent an entire landscape. In fact its undulating musculature has been likened to Uluru/Ayers Rock. Recently, however, such hedonism has been overshadowed by environmental pollution and health hazards. The idyllic deserted beach of cigarette advertising now has to compete with images of oil-slicked animals and syringes in the sand.

It is this vision of 'paradise lost' that Pam Debenham has worked into her 'irradiated' Hawaiian shirt poster. Island resort names have been replaced by nuclear test sites, 'Marshall Is, Muraroa Atoll, Bikini', and lurid mushroom clouds puncture the endless blue. She thus returns us to the image described in the 1950s novel and film *On the Beach*.

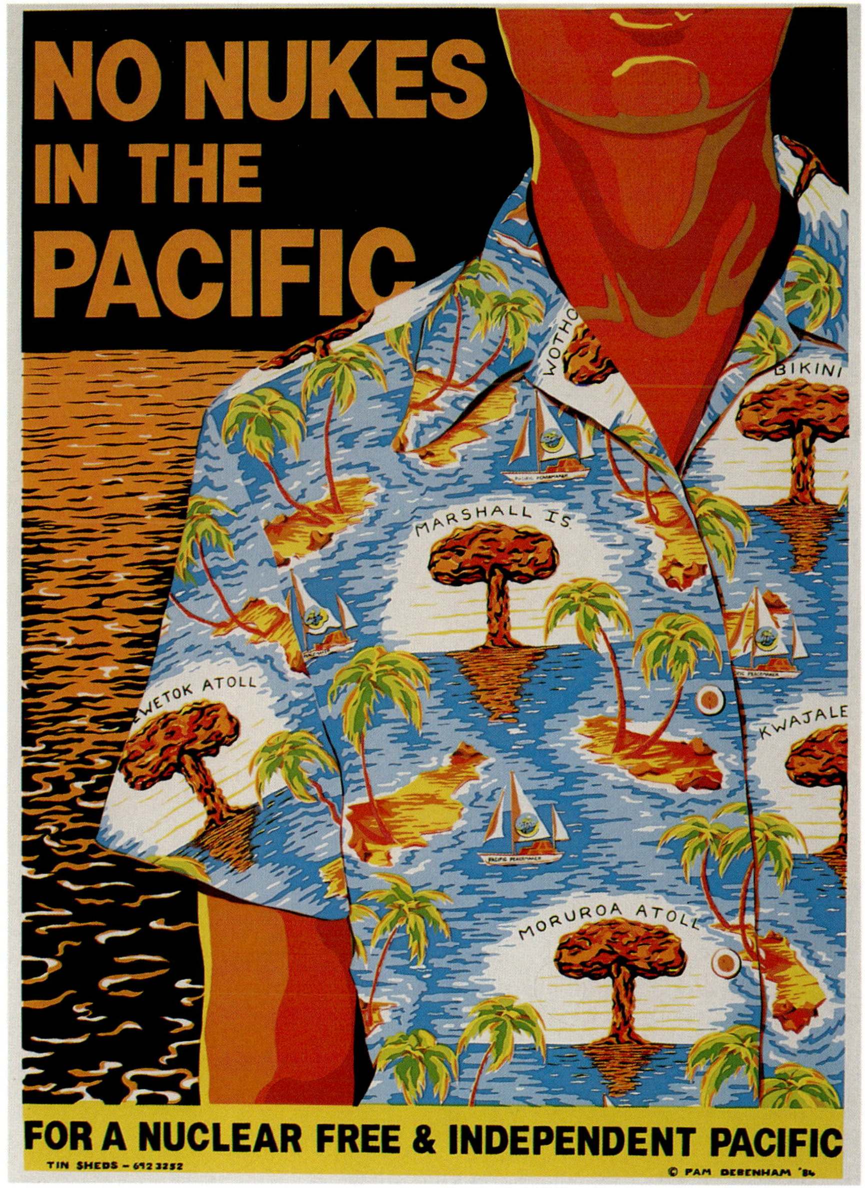

No Nukes in the Pacific, **1984 'Irradiated' Hawaiian shirt poster by Pam Debenham**
Exhibit no. 17

FROM WILDERNESS TO NATURE STRIP — Ann Stephen

SUBURBAN REAL ESTATE

'Our plate-glass picture windows,
Venetians open wide,
In the land where nothing happens
There's nothing much to hide.'

Edna Everage, *All Things Bright and Beautiful, 1968*

The suburbs, where most people live out their lives, are least subject to picturesque or sublime treatment. Instead they are linguistically captured by real estate agents' hyperbole, and the resilient line of suburban satire delivered by the likes of Edna Everage, late of Moonee Ponds. It is this tradition, sometimes surreal, sometimes pop, that artists have mined for textile designs, tea-towels, T-shirts, posters, and post-cards. They have focused on the signs of difference in the face of the uniformity of the quarter acre block — the customised hedge or letter box, the wrought iron-work, the family dog, the garden gnome, or the tyre swan.

Those locked inside the sprawling suburbs, without the cash or wheels to reach the beach, the mountains, or the city, have been given some voice in recent years through community arts centres, like Garage Graphix in the west of Sydney.

PASTORAL 'HISTORYLAND'

'...draw a wire fence and a few ragged gums, and add some scattered sheep running away from the train. Then you'll have the bush all along the New South Wales Western line from Bathurst on.'

Henry Lawson, *In a Dry Season*, 1892

Henry Lawson was a childhood victim of an earlier pastoral industry gone wrong and knew that the pastoral landscape could be represented by the most economical of signs.

The dominance of the pastoral landscape tradition in early twentieth-century Australian culture was displaced in the 1940s as desert and suburbs encroached. Today, with an economy no longer riding on the sheep's back, farmed land enters the mass media as either rural crisis or nostalgia.

However, other industries which are remote from an agrarian base, such as car manufacturers, are still happy to use pastoral settings in their advertising, because they provide such a contrast to the product being sold. Take the Ford Corsair family who have stopped to inspect a roadside 'colonial heritage village'. The vaselined, brown backdrop makes it clear that this is 'historyland'. No one appears to live or work here but, it is a quaint spot for city folk to stretch their legs before hopping back into their sleek new wagon.

RAINFOREST

'...the wilderness we now value and try to protect came with us, the invaders. It came in our heads, and it gradually rose out of the ground to meet us.'

Eric Rolls, *A Million Wild Acres*, 1981

In 1788 rainforests ran the length of the east coast of Australia and were sites of fascination and fear for the colonists. They provided valuable timbers and visual symbols of the new land, but stood as impediments to development of rich pastoral spreads. 'Let us regard the forests as an inheritance, given to us by nature, not to be despoiled or devastated, but to be wisely used, reverently hon-oured and carefully maintained'. This warning by the botanist Baron von Mueller was unfortunately ignored for over a century.

Modern mass media campaigns showing pictures of devastated forests alongside apparently 'virgin' wilderness have led to rainforest areas being valued for their wilderness as much as for their ecological diversity. Significantly, in the now very popular genre of rainforest photography, there is seldom any sign of how forests have been modified by humans, whether Aboriginal people or post-contact Europeans.

THE SPACE LEFT OUT (THE BACK)

'This is ther Never-Never — ther lars' place Gord made.'

Barbara Baynton, *Billy Skywonkie*, 1896

Of all the landscape sites the 'out-back' is most contingent on the relation-ship between European settlement and the space left out (the back). The domi-nating feature of this imagined space is the dead/red heart of the Centre which exerts a powerful gravitational attraction upon the vast, undifferentiated desert. So much so that the rock, Uluru/Ayers, which is at its semi-mythic core has become the most exported of Australian landscape images.

Aboriginal people have long been depicted as being no more than just another part of outback flora and fauna, yet their own picturing of the desert comes from the inside out. Their images refer to an intimately known land, rather than to a *Mad Max* vision of a vast empty desert.

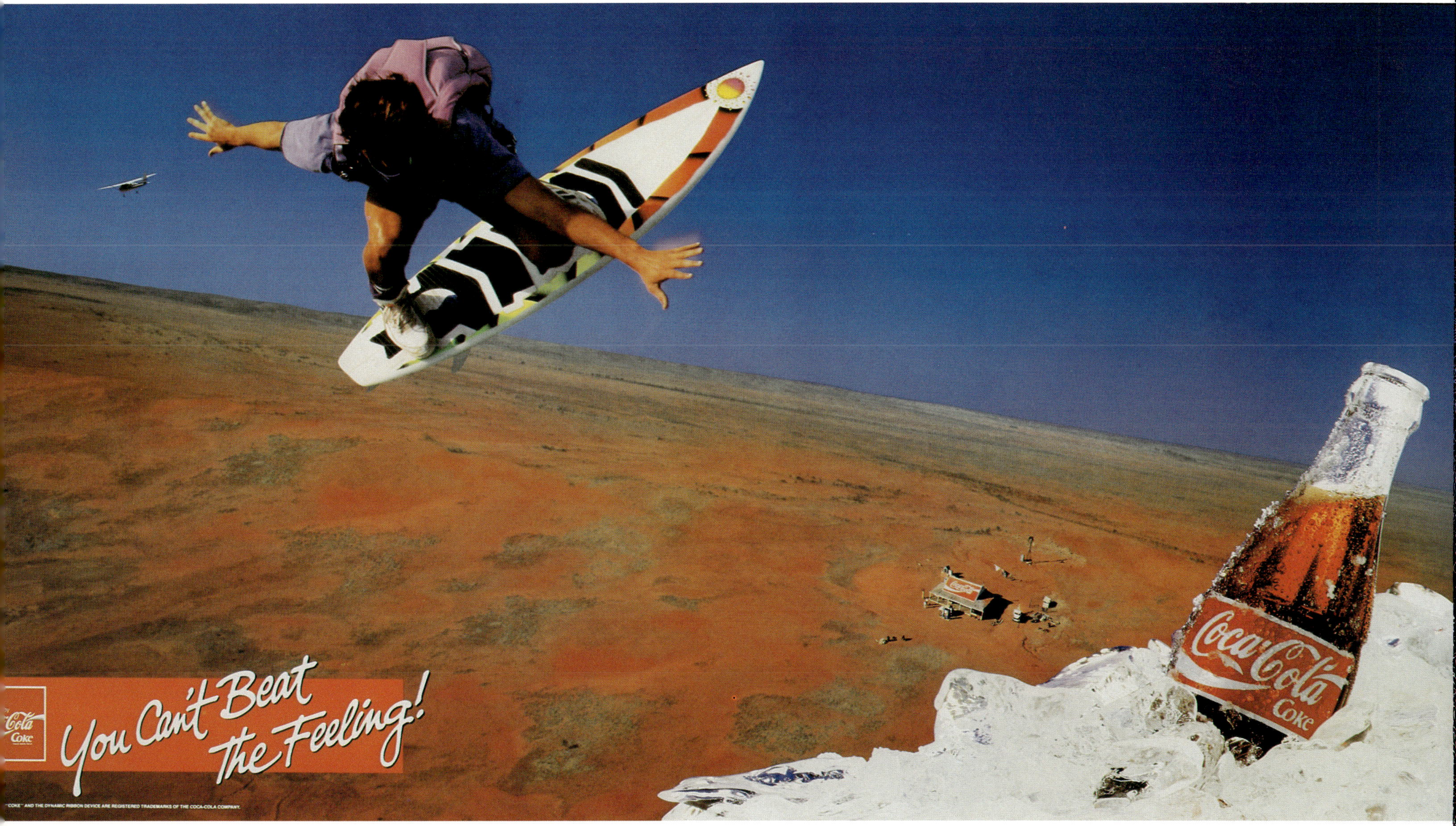

You Can't Beat the Feeling, 1991 Aerial view of desert country featuring skydiving surfies
Copyright: The Coca Cola Company Exhibit no. 55

FROM WILDERNESS TO NATURE STRIP — Ann Stephen

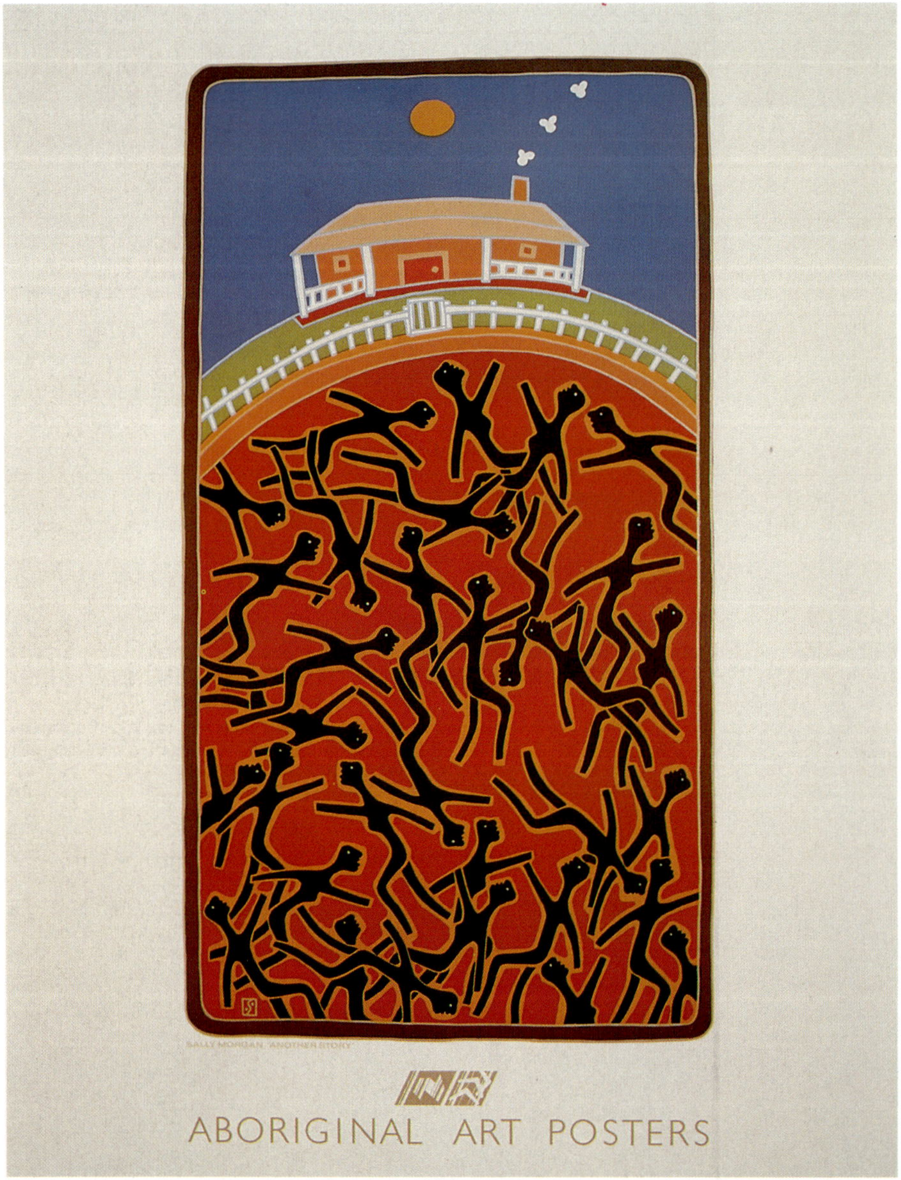

***Another Story*, 1988 Designed for the Australian Bicentennial Authority by Sally Morgan**
Exhibit no. 39

ABORIGINAL PRESENCE

'Land is the generation point of existence... It is a living place made up of sky, rivers, trees, the wind, the sand, and yet it is not a thing. It is a living entity. It belongs to me. I belong to the land. I rest in it. I come from there.'

Pat Dodson, *Aborigines: a Statement of Concern*, 1980

The shadow cast across all these landscapes is the Aboriginal presence (or present absence). As the anthropologist W. E. H. Stanner observed in 1968, 'Every fence in Australia encloses land that was once the sole or the shared possession of a particular group of Aborigines'. This is the subject of Sally Morgan's print, *Another Story,* with its restless black bodies under the soil seeming to press up on the pastoral vision that has been laid over them. Such imagery is scarcely surprising, as it was designed by an Aboriginal artist for the 1988 Bicentennial Authority.

What comes as a shock, however, is to see an aerial view of desert country bearing an uncanny resemblance to the patterns in dot paintings, in a 1991 Coca Cola campaign featuring skydiving surfies (see previous page). While Aboriginal people occupy a central role in outback imagery, with the desert as the pre-eminent landscape for Aboriginality, Koorie and Murri imagery has, in the last decade, begun to depict a contemporary urban and suburban space.

Aboriginal visual cultures are beginning to transform the way land in Australia is seen and understood. In particular, abstract designs that speak of the 'land', representing specific Aboriginal Dreaming Places, have achieved popular currency equivalent to the status that Namatjira held for a previous generation. We can see these designs, now mass-produced on posters and textiles, as a means of communication, developed and controlled by the artists, to speak to non-Aboriginals about their attachment to the land. That their meanings cannot be fully understood by the non-initiated has increased the status of the relationship Aboriginals have to the land and mystified the 'landscape' for non-Aboriginals.

In Australia, our landscapes continue to haunt and shape our ideas about national identity. **The Lie of the Land** explores the process of change that landscape imagery has undergone in the last two decades. Our relationship to the land has assumed a sense of urgency as we realise that the earth's future will be determined by the environments we have shaped.

RULES & ADVICE FOR NON-ABORIGINAL PEOPLES ENTERING ABORIGINAL LAND — Terry Sams

To enter Aboriginal Land, first obtain a written permit from the local Land Council. Secondly, contact the person in charge of the community, by telephoning the council office, and making sure it is alright to go there. You must give two to three weeks notice before you arrive.

Ensure that there are not any ceremonies on, such as:

- Funerals;
- Initiation ceremonies;
- Circumcision ceremonies for boys;
- Women's ceremonies;
- Smoking of the house where the deceased used to live and burning and burying of the deceased's belongings.

It is wise to telephone again on the day, just to make sure that it is still alright to visit.

If, when you are at the community, you wish to go out somewhere, please ask or take someone with you. It is much easier to do this than to get into trouble by going to a sacred site or to land which belongs to someone from another community. It is important to ask the traditional owners for permission before you visit their land.

Aboriginal people will not look you straight in the eye when you are talking to them. Europeans think this is very rude, but in Aboriginal society it is good manners and is the correct way.

When you see a group of people sitting down and you wish to speak to someone in that group, please do not stand and talk. You must sit down to talk to whoever you wish to speak to.

When one is given a gift, one does not say anything, but sometime later will give something to the person who gave the gift. There are no words for please or thank you.

When someone has recently died, it is forbidden to say their name. By speaking their name, that person's spirit will haunt you forever and may affect your mind.

I hope that this report will be useful for all non-Aboriginal people who are entering Aboriginal Land.

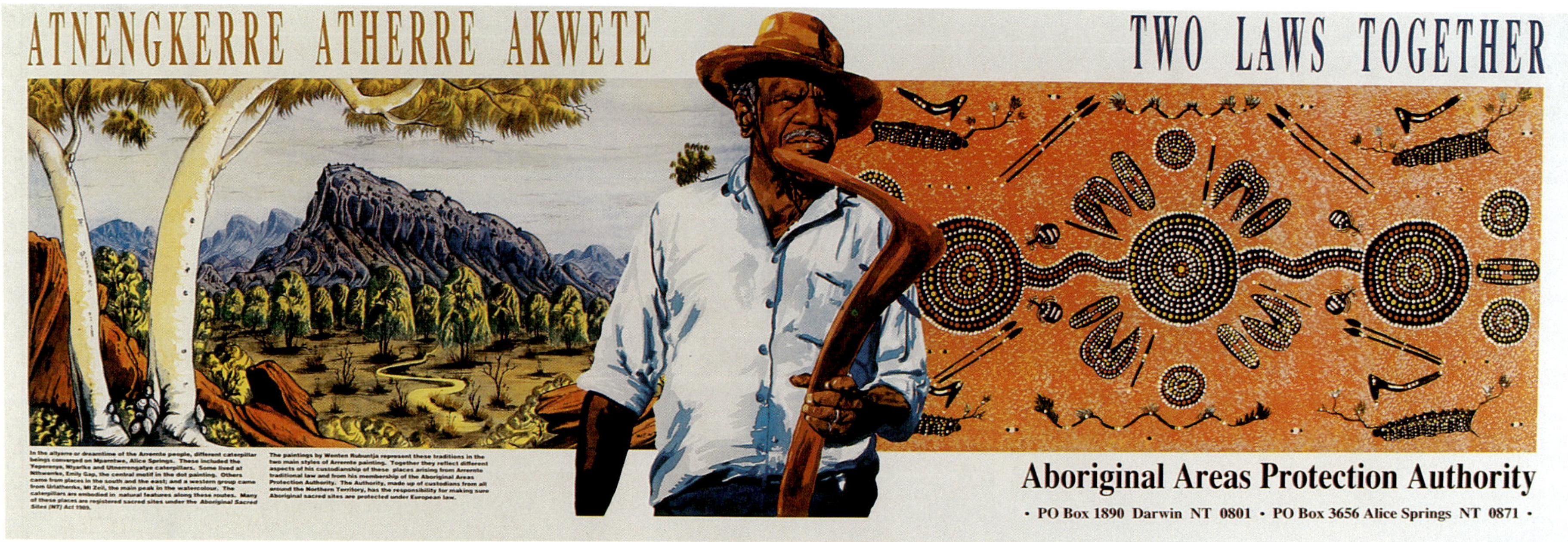

Two Laws Together, 1991 **Designed by Wenten Rubuntja & Chips Mackinolty for the Aboriginal Areas Protection Authority** **Exhibit no. 60**

THE LAND IT BELONGS TO THE LARAKIA PEOPLE — Marjorie Bilbil

We went to the meetings with the Larakia people.
The land it belongs to the Larakia people.
Anyway we had this big day with the elders and others,
Who are living at Belyuen, showing the site and hunting places.
We went to the islands
And the elders told the story of the lands
And the Dreaming where they were,
And told people not to go near it
Because it might drown the boat.
There were sites on the land as well.

The people at Belyuen was talking to the judge
About the land
And what the tribe was,
Where they came from,
What their language was,
Where your Mother and Father came from,
And what language they spoke.

Even I spoke to the judge.
He asked me
where do I come from,
what language I speak,
and what tribe I come from.
He asked me
about the story that I heard from the older people
about the land.

What I heard from the older people was important,
for the land was the site of the Dreaming
and the younger people are still learning.
The older people say the Dreaming
and the sites are very important to us all.
The site and the Dreaming should be protected very well.
The elders say the Dreaming is dangerous
some things might happen to us all.
That's why it is very important.

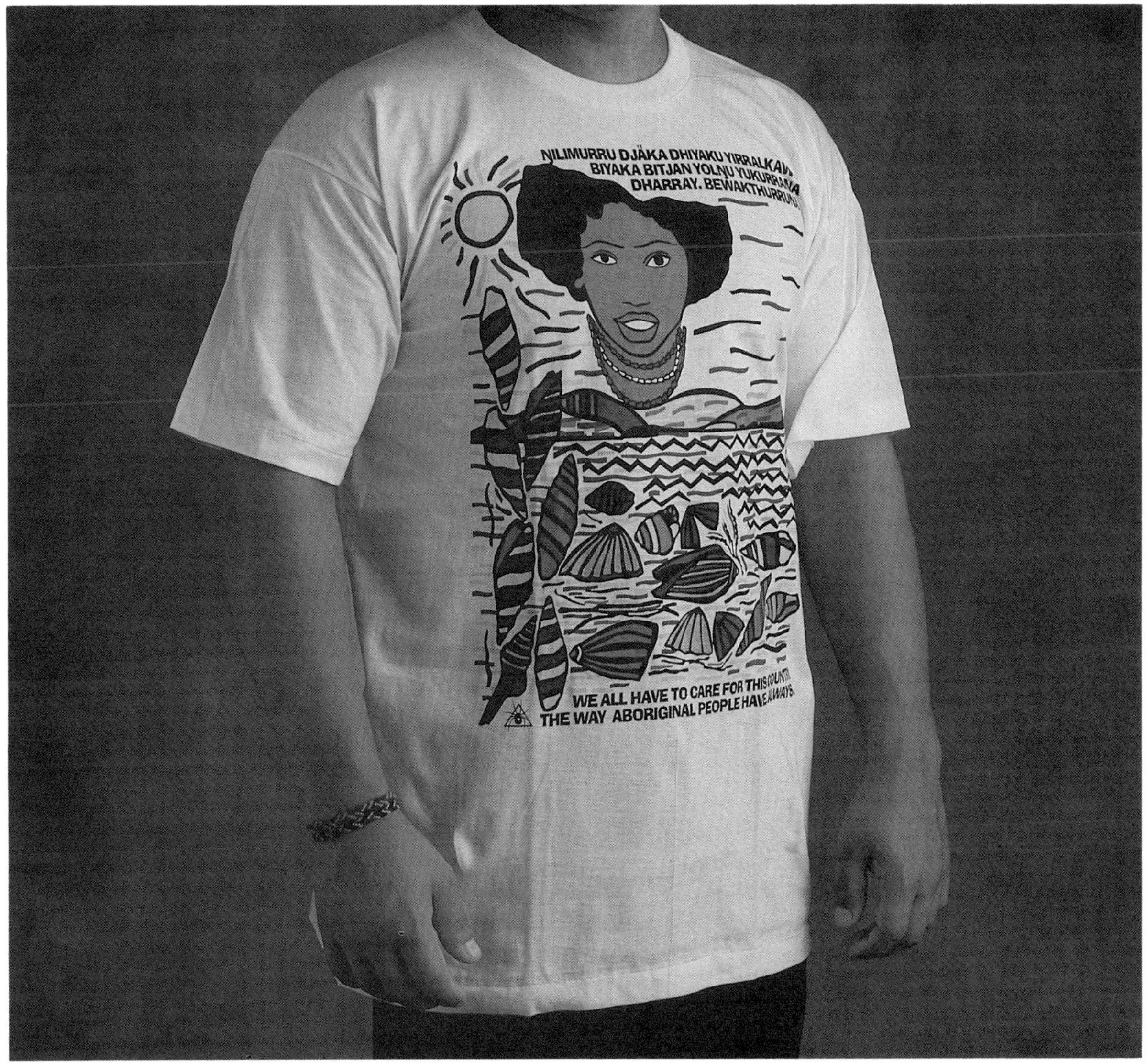

Care for this Country, 1988 **Designed by Alison Alder** **Exhibit no. 71**

LICK A LANDSCAPE: PLACES ON POSTAGE STAMPS — Richard Peck

Not until the 1970s were stamps depicting natural landscapes issued in Australia. During this time, multicolour printing revolutionised stamp production, increasing the range of imagery which could be used on such small objects. Concurrently, governments considered the traditional depiction of the monarch as less relevant than the opportunity presented by stamps to convey images and messages of economic, political, and social content.

The stamp and postcard images in this exhibition display both the variety of the Australian landscape and the differing ways it has been depicted since 1970. Compare the 1976 'Australian Scenes' stamps with the 1988 'Panorama of Australia' issue. The former was produced by the photogravure process, which meant that the artist had to produce a separate piece of artwork for each colour of the stamp. Now modern, computer-assisted colour separation and offset lithography is used, and there is an infinite variety of possible photographic images. Hence, the 1988 issue looks just like photographs or postcards.

The pastoral image of sheep has been depicted on Australian stamps for a longer time than the Australian landscape. A prize merino ram was shown on a commemorative set of three stamps of the same design issued in 1934 for the centenary of John Macarthur, one of Australia's pioneer wool producers. An adapted version was used for the normal five pence stamp, as part of the animal and bird series of 1937–1952. Again, these stamps were produced in one colour only.

Three sheep were shown on a small 25-cent stamp issued in 1969 depicting primary industries. A scientist with a lamb is on an 18-cent stamp from 1971 as part of a group of stamps showing the relationship between animals and humankind, issued for the Centenary of the Royal Society for the Prevention of Cruelty to Animals. *Shearing the Rams* by Tom Roberts, a $4-stamp from 1974, was part of a high-value series depicting Australian paintings. By 1989 it was time to depict sheep again. Noting the previous designs, the designer's brief was to show sheep 'in a heroic mode'.

ABORIGINAL INFLUENCE

The perception of Aboriginal art has changed over the years. A commemorative set for Victoria's Centenary in 1934 depicted an Aboriginal person gazing at Melbourne. The only other Aboriginal person to be shown on a stamp was 'One Pound Jimmy', who appeared on eight pence, halfpenny, two shillings, and six pence stamps used between 1950 and 1966. He got his name from his habitual answer to the question of how much he would charge for a job. His stock response was 'one pound boss'. Although it is a tradition that no living person, except the monarch, be portrayed on Australian stamps, Jimmy was in fact alive when the stamps were issued.

The 1971 'Aboriginal Art' stamps, commissioned by Australia Post, were the work of a non-Aboriginal designer John Mason. In 1988, however, Ray Meeks, an Aboriginal artist, was commissioned to design the 'Australia Day' stamp. The design is based upon images of two Aboriginal spirit beings, the human-like Wanjina and the Rainbow Serpent, who were responsible for the creation of the earth and land including every physical feature and all living things upon it. The design is further enhanced by using elements from the art styles of Arnhem Land and Central Australia: the X-ray technique, cross-hatching, and dots.

The 1988 'Art of the Desert' series was designed by a non-Aboriginal graphic artist from Melbourne, Janet Boschen, and was based on Aboriginal paintings from Central Australia. Some Aboriginal stamps cannot be shown. The first series for the Australian Bicentennial issued in 1984 depicts rock paintings by the first Australians. During the design stage of this series, it was discovered that one of the artists had used sacred images and so this design could never be used.

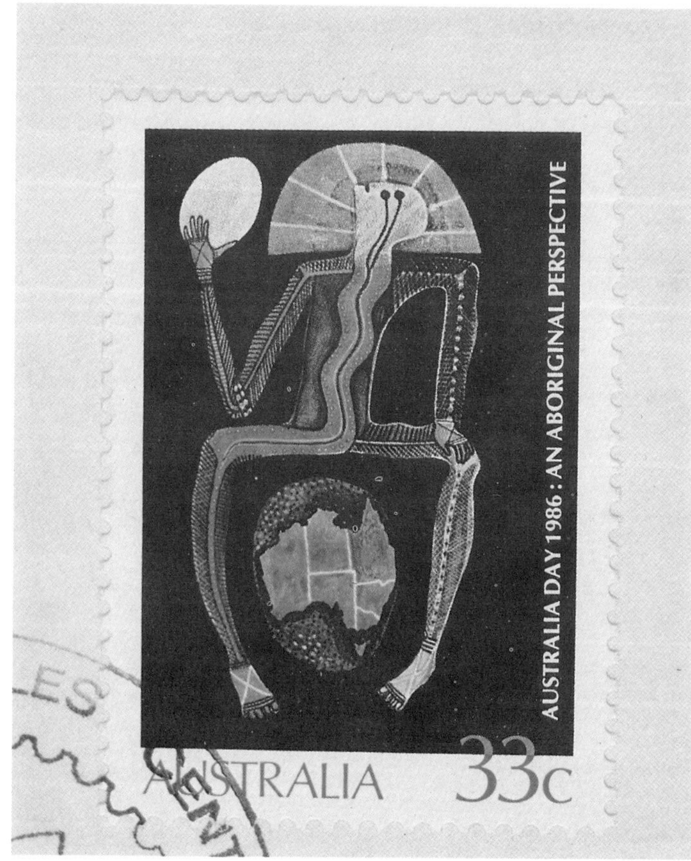

Australia Day 1986: An Aboriginal Perspective
Copyright: Australia Post

While stamps for business mail may one day be replaced by electronic methods of communication, personal letters and postcards will still employ these miniature ambassadors of their country. It is also certain that the popularity of landscape imagery will ensure future Australians continue to lick a landscape.

CAMERA NATURA: LANDSCAPE IN AUSTRALIAN FEATURE FILMS — Ross Gibson

Mad Max fights for hegemony over the land, while picnickers are subsumed into it, never to return, and the man from Snowy River spurs his small and weedy beast in a race to master it. The land also maps out the sorrow of the stoic shearer's wife who sees cause to lament that on Sunday, Jack Thompson is too far away.

The common denominator in all these situations, the motif and ubiquitous central character, is the Australian landscape. Its foregrounding (plus midgrounding and backgrounding) is such that the country becomes something much more significant than the environmental setting for indigenous narratives. In so many ways these films are *about* the Australian landscape. Cinematographers — Boyd, McAlpine, Hansen — become *auteurs,* as infinite space, light, and colour are combined to elicit audience gasps at sublimity. It is *our* beautiful country. Consider *Plains of Heaven, My Brilliant Career, The Chant of Jimmie Blacksmith, Journey Among Women,* or the early stages of *Gallipoli.*

Why were mainstream film-makers, audiences, and critics in Australia under the spell of some land spirit throughout the 1970s and 1980s? Why this obsession with the natural environment? What can this preoccupation tell us about our culture — cinematically and generally?

This reverence of the landscape is clearly the result of an alienated society's experience around the ridges of a vast, underpopulated, and speciously indomitable country. But what of the effects? The legends of the awesome land imply the negligibility of social action. The society cannot be seen to be directing the environment in its own interests (despite the multi-national advertisements about their quiet achievements in the ecosystem). It cannot stretch out to cover all the unsubdued continent. In response to this implicit futility, the more heroic myths of the individual versus the environment fill the screen (*Mad Max, Sunday Too Far Away, The Irishman* . . .). Society en masse cannot make a mark on the land, so individuals set about carving niches for themselves, communing with the spirit of the land, reading its messages. In the inhuman landscape humanity flourishes.

It is through the rendition of Australian landscape as sublime and suprasocial that film-makers are attempting to create Australian culture even as, paradoxically, they are promoting the view that a society cannot really make much of an impression on such a habitat. Or perhaps they are attempting to read some innate Australianness into the landscape as they aim to stamp their own inner natures on the external nature of the continent.

It is not by chance that a crop of extensively publicised and acclaimed films highlighting a unique Australian landscape came to light during the decades of resurgent nationalism. The landscape cinema asserts an Australian difference. The films say, 'Here is a key to our identity'. (Certainly this is how they were marketed both at home and abroad.) Just when it seemed that the cultural cringe was obsolescent, a burgeoning film industry came along to display and participate in Australia's historical adolescence.

One reason for the long list of Australian colonial costume dramas derives from the fact that the time-setting allows the film-makers to focus clearly on the theme of individuals, or isolated groups of 'primitives', in confrontation with nature. More recently *Mad Max, The Last Wave,* and *Plains of Heaven* have demonstrated that nineteenth-century themes can be transposed with contemporary themes. The landscape still seems to hold the key to Australian mythologies. However, as accountants and many reviewers would agree, there is no point in complicating things: put those 'eternal' themes in a setting that shows them off with 100 per cent clarity.

THE VICIOUS CIRCUIT

Under Australian skies we are looking with Western eyes and the contradiction is manifest in the films. The Australian film industry subscribes increasingly to the 'ethic' of investment entertainment and the 'Hollywood' narrative format is endemic. Generally, it features a naturalistic, single-theme story which purports to concentrate on the fortunes of intimately known individuals acting and reacting in a 'real' universe. It sounds like an uncomplicated formula for success. But in Australia the not-quite-different filmic landscape is always there, claiming priority of place as the central character. Australian nature resists colonisation by dominant film discourses. Yet, even as the film-makers are rejoicing in this, and saying 'this is where we are different — look at that landscape — it's our unique country', the Western culture is refusing to allow a distinctive nature to imprint itself on Australian film form. The film-makers are obeying all those rules of dimension and ocular clarity which inscribe the images of the continent with traditional European attitudes about landscape art, attitudes which have both grown out of and bolstered a specifically European view of nature and culture. A continent which has spent eons in its own geographical and cultural locales is promptly framed within European parameters. (See the Manet-like tableaux in *Picnic at Hanging Rock*.) In yet another way the Antipodes become a reflection of Europe.

The search for Australian specificity is one of the motivating chimeras of Australian cinema, drawing thoughts and gazes centripetally into the outback. Of course, white Australians will never arrive at anything other than a predominantly white Australian ethos. But in a necessary process of self-definition they habitually seek more 'native' vision, waiting for the land of the dreamtime to stamp itself on the culture.

This is an edited version of an article which first appeared in *On the Beach*, No. 1, August 1983.

THE ROAD THROUGH THE DESERT: OZ VIDEO CLIPS — Kathleen Phillips

Desert descriptions can never be seen as innocent of cultural reference. 'The desert' is always a pre-existing pile of texts and documents, fantasies, legends, jokes and other people's memories, a vast imaginary hinterland which most coastal dwellers like to dream all their lives of seeing in reality one day. (Meaghan Morris 'Two Types of Photography Criticism', *The Pirate's Fiancee*, Verso, London, 1988)

We are all exposed to music video clips, whether we like them or not; on television, in bars, and even in museums. One thing that sets Australian music videos apart from their European and American counterparts is the way that producers often choose to work outside the studio. In many clips, producers try to draw connections between Australian culture and the environment. Although the urban environment is often featured, this article focuses on how music videos use non-urban landscapes, in particular the desert landscape.

The video for *Beds are Burning* by Midnight Oil was produced by Tony Leach in 1989. It features spectacular and thoughtful use of Australian landscape and was a welcome relief from the flood of Bicentennial imagery that had dominated the previous year. But what does it mean, this Land Rover trailing across western deserts to the streets of Redfern? Do the millions of listeners around the world, who know and love this song, realise that it is about land rights?

All power to Midnight Oil for saturating their lyrics with politics, but what about the images of the desert, empty but for the band? This is the romanticised representation of the desert favoured by film-makers: the uninhabited, endless, unspoilt land. While their intention may be different, we can draw comparisons between the boys thundering through the outback, guitars in hand, with the work of film-maker Charles Chauvel.

Charles Chauvel was a nationalist. In the opening titles of *Forty Thousand Horsemen* (1941) he describes Australia as a 'land won by conquering people toughened by adversity'. According to film critic Sandra Hall:

Chauvel was the first film-maker to be attracted by the dramatic force of the Australian landscape, (which) makes his . . . films . . . remarkable . . . for their haunting sense of place and their largeness of spirit.

Film historian Stuart Cunningham has written of Chauvel's attempts 'to capture the Australian spirit embodied in landscape', and that is what Midnight Oil aimed to do in *Beds are Burning*. It is a different Australian spirit now. In Chauvel's time, humankind had to *conquer* the spirit of the land, to overcome the droughts, fires, and floods. Now we have to *work* with that spirit. In this video clip the spirit is represented (not very subtly) by the painted Aboriginal dancers superimposed over the images of landscape. The message is clear — cooperation with the Aboriginal community and with the land, droughts, fires, floods, and all.

The crucial difference between the work of Charles Chauvel and that of Midnight Oil is in their attitudes. Chauvel's attitude to the country is one of post-colonial possession and ownership, while the latter's is one of sympathy toward both the traditional owners and the land itself.

The intentions of Midnight Oil are admirable, but the images of the band alone in the desert do raise some questions. Is *Beds are Burning* a land rights statement or a means to sell records? Or, less cynically, is the song exploiting nationalist mythology for a political purpose?

DESERT STYLE

In 1982 record producer Martin Fabinyi established the genre of desert music video clips with *Great Southern Land* by Icehouse. No politics here: the desert is pure style, a nice backdrop of texture and colour. Images of burning bushes, Iva Davies in a leather jacket climbing a rock, no sign of life other than a prehistoric-looking lizard.

With lyrics about primitive man and the ghost of time, the message is clear — life and the universe started here. The featured Rock has assumed significance as the manifestation of 'the Spirit of the Land', the *real* Australia. The burning bushes (or are they burning crosses?) serve to reinforce the spiritual nuance of the clip.

There is something mysterious about the desert. As Meaghan Morris says:

It is a space accorded the status of a reservoir of places where nothing might be, or anything might happen. In traditional legend, birds fly backwards there; rivers run against nature, the sand spawns fishes, and inland seas are lapping just beyond the dry horizon. In contemporary speculation, lost creatures reappear and vanish; regions belong obscurely to alien powers, while the wasteland hides a monstrous proliferation of caverns breeding new forms of experimental warfare.

No wonder Iva is wearing his leather jacket !

INXS, an Australian band which has received international acclaim, has also used the landscape and dramatic camera work to enhance its 'Australianness'. The video clip of *Kiss the Dirt* (1986) is a direct descendant of *Great Southern Land*. As well as the *de rigueur* lone figure in the empty landscape, this clip has the intrusion of the road, traffic, trains, and lines of communication.

This leads to the 'road movie' genre of video clips (*Wide Open Road,* The Triffids; *Burn for You*, INXS) in which country roads and highways are frequently featured. In some ways, this is part of the same myth: that the heart of Australia is found outside the major cities. The view from the highway is what most Australians see of the local countryside. This time the road and small towns combine with the unpopulated outback to represent the real Australia.

CULTURAL WASTELAND

In the Warumpi Band's *My Island Home* (1987) the desert is seen from a different perspective. Instead of a romantic view of an infinite, empty landscape, we experience the desolation and isolation. We experience the desert through the eyes of a coastal dweller, a man who is forced to live in the desert and can only dream of the sea.

The majority of Australians live around coastal areas rather than in the desert. According to Meaghan Morris, this is quite consistent with the myth of the *real* Australia:

The enduring seductiveness of the myth resides in the reversibility of its meanings. The inland is also repulsive; an image of a natural dreariness, dullness, desolation and monotony which slowly encroaches upon settlements, and spreads its emptiness by contagion to culture. For a century, recurring projects prove that Australia is or is not a cultural desert; poor towns, and unpleasant parts of cities, are not jungles here — but deserts.

The anomaly about the preponderance of desert in our music clips is that Australia's rock music culture thrives in our major cities. Outside the cities it is Slim Dusty and Reg Lindsay who rule the airwaves.

How does the city audience feel about the bush in video clips? Meaghan Morris believes that 'in urban imaginations, that space is *there* — immense, unique, invested with meaning, and rather expensive to tour'.

She goes on to explain how entrepreneurs have seen a way to make a dollar:

In the Eastern cities, there is a whole cultural industry for promoting the presence of the desert. Television documentaries, postcards, films of the lavish landscape, celebrations of books of Australiana, even ecological campaigns keep accumulating and circulating images of wondrous far horizons — which function as so many signposts to another reality, a *real* Australia.

Whatever the theme, love, land rights, or loneliness, whatever the medium, the image of the desert lends strength to the message.

Based on a talk given by Kathleen Phillips at 'Constructing a Culture' conference, Melbourne 1990.

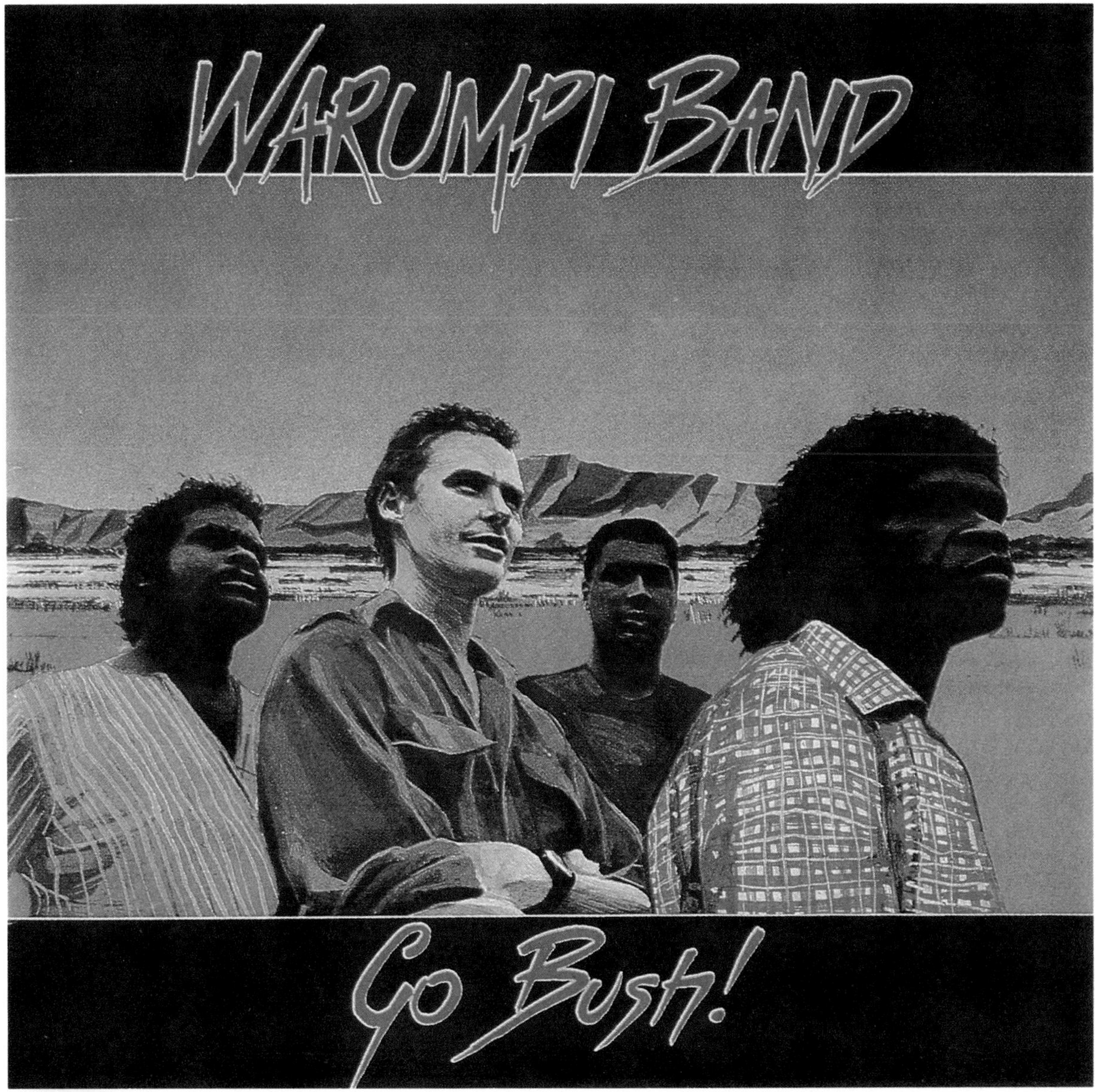

Go Bush Cover of Warumpi Band's 1987 album Designed by Janice Barkla from photograph by Mark Manion Artwork by Mark Skarski & Karen Clarke Copyright: Trafalgar Music

THROUGH THE DISTORTED GLASS — Annette Shiell

Snowdome, snowshaker, snowscene, snowstorm, blizzard ball, blizzard weight, waterball, waterglobe. The name may have changed, but these magical miniature worlds have been a source of fascination since the late 1800s.

Credit for the creation of the snowdome must be given to the French. The first known waterglobes were an extension of the elegant glass paper weights that French glass artisans of the nineteenth century were renowned for. In 1878, the *Reports of the United States Commissioners to the Paris Universal Exposition* noted that 'French exhibitors of decorated glass... exhibited... paper weights of hollow balls filled with water [and] a white powder which, when the paper weight is turned upside down, falls in the imitation of a snowstorm ...'. (M. McMichael *Snowdomes*, Abbeville Press Inc., New York, 1990)

Several years later, the French again led the way in snowdome development. At the 1889 Exposition Universalle in Paris, an enterprising manufacturer commemorated the construction of the Eiffel Tower by developing and selling the first souvenir snowdome. This landmark snowdome consisted of a ceramic model of the Tower, encased in a water-filled glass sphere, which was mounted on a square ceramic base.

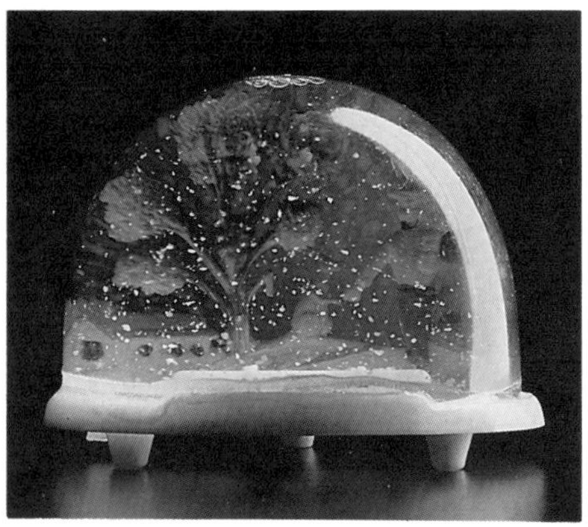

Australia
Encapsulates sentimental images of a halcyon past Exhibit no. 112

Snowdomes soon caught the imagination of the public and by the late nineteenth century these waterglobes were being produced simultaneously in France, Czechoslovakia, Austria, Poland and Germany.

During the 1920s, German firms began to export waterglobes to the United States and Canada. These early designs consisted of spherical globes, mounted on cobalt blue bases on which the names of local towns were inscribed. This tradition was borrowed from the nineteenth-century German pottery and china manufacturers who 'souvenired' their products with local scenes and inscriptions.

In 1927, the first American snowdome was patented when Joseph Garaja of Pittsburgh, Pennsylvania filed an application for a glass waterglobe of 'artistic attractiveness and novel ornamentation'. His designs proved very popular and before the year was out they had been copied by enterprising Japanese manufacturers.

The turning point in the development of the snowdome occurred in the 1950s. In West Germany, new plastic materials were used to create an innovation in snowdome design: the flat-bottomed, oval-shaped model. One of the German manufacturers claimed the design was inspired by the domed view of a winter landscape seen through the rear window of his Volkswagen beetle car. This plastic oval-shaped model has since dominated the snowdome market, as is evidenced by the collection in the **Lie of the Land** or a perusal of any souvenir stand, and has now become synonymous with the word snowdome.

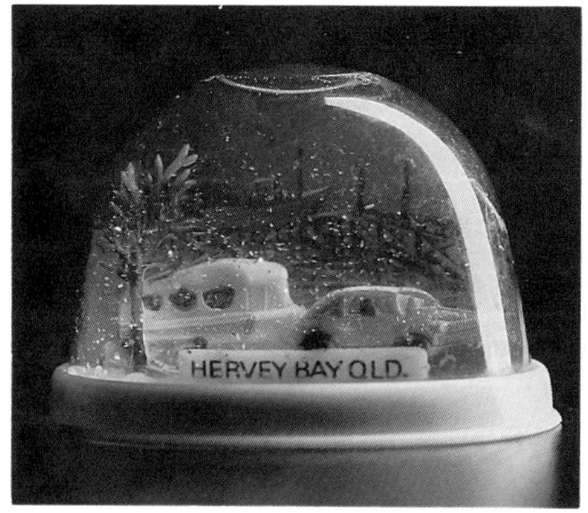

Hervey Bay, QLD
Immortalises the great Aussie family holiday Exhibit no. 94

Since their creation, the contents and consistency of snowdomes have been a source of constant vexation for manufacturers. Over the years, they have experimented with all manner of materials in their quest for white fluffy snow, namely: bone chips; coarse pottery flakes; fine fragments of porcelain and china; sand; sawdust; mothball flakes; ground raw rice; camphor/wax; minerals such as meerschaum; and eventually finely chopped plastic. An Australian collector, who prefers to remain nameless, even resorted to dessicated coconut as a snow filler — with disastrous mouldy consequences!

Assembling snowdomes has proved to be equally exacting. Originally they were assembled completely under water to ensure a tight seal. In the 1950s, a firm of Austrian manufacturers added antifreeze to the water in their snowdomes to prevent them from continued freezing and cracking. At one stage, an embargo was placed on the import of Asian domes to the United States when it was discovered they were filled with polluted water straight from Hong Kong Harbour.

Today, manufacturers generally follow one of two approaches. They either add chemicals such as glycol to the water which acts as an anti-freeze

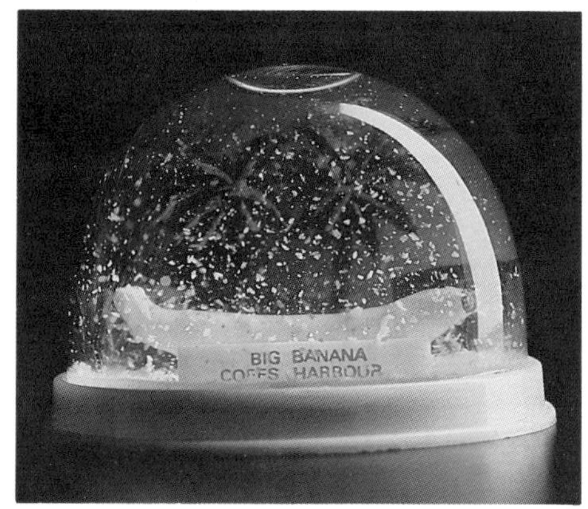

Big Banana, Coffs Harbour
The anomaly of this tropical landmark under snow is unquestioned Exhibit no. 91

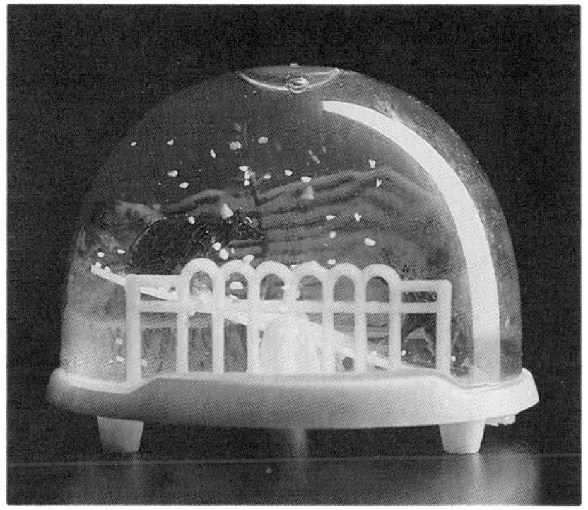

Tasmanian Tigers
These same 'tigers' are sold as souvenirs of the US National Parks Exhibit no. 134

while purifying and thickening the water, or they ship the snowdomes without water and include instructions for filling.

UNIVERSAL APPEAL

The appeal of snowdomes is difficult to define. With their distorted shapes, garish colours, coarse brushwork and incongruous snow, these quirky *objects d'art* border on the kitsch, but conversely these very same elements inexplicably combine to give them a magical quality. Snowdomes cross all barriers and serve as happy reminders of past experiences and childhood adventures. In the 1941 film classic *Citizen Kane*, a snowdome is the key to the mystery of the media tycoon's dying word. By shaking the snowdome, Kane is carried back in time, to the happy memories of his lost childhood.

Snowdomes are miniature utopian worlds. They encapsulate the very heart of the landscape sites portrayed — whether that heart be real or a romanticised interpretation. In the **Lie of the Land**, one of the pastoral snowdomes, *Australia,* conjures up images of a halcyon past where swagmen and kangaroos were at one with the bush. *Surfing Lakes Entrance* (see illustration on the front cover of this catalogue) focuses on Australian beach culture, embodying the youthful life and excitement of carefree summer holidays. *Hervey Bay QLD* immortalises the happy family holiday with mum, dad, and the two kids arriving in a reliable sedan with the caravan in tow.

Through the distorted glass, our world becomes not only rose-coloured, but expanded beyond reality. In a snowdome, the incongruity of Australian landmarks such as Uluru, the Barrier Reef and the Big Banana being covered in snow is glossed over. The anomaly of having these hot/tropical landscape sites surrounded by a snowstorm elicits nothing more than a fond smile of acceptance. In one of the Tasmanian snowdomes, the knowledge that the Tasmanian Tigers have an uncanny resemblance to the brown bears of America and are riding on a seesaw, does nothing to diminish the charm of this Australian pastoral scene.

In fact, the more bizarre they are, the greater their appeal seems to be. The snowdome entitled *Kangaroo*, depicting a pastoral scene in which a kangaroo is jumping through distinctly European fir trees amid a snowstorm inside a television-shaped salt and pepper shaker, is perhaps the *piece de resistance* of the **Lie of the Land's** snowdome display.

Snowdome collectors delight in finding that special snowdome — the one with the mistake. Errors caused by fast mass production such as incorrect name plaques, spelling errors and upside down figurines make these highly sought-after collectors' items. According to one Australian collector, the seemingly straightforward *Parliament House Canberra* snowdome is quite a collectible. Reputedly, the first snowdome produced was withdrawn from sale when it was discovered that the flag fluttering behind Parliament House looked decidedly more Japanese than Australian.

At first glance, snowdomes appear to be in the same league as any other mass-produced souvenir item, in that they are portable, affordable keepsakes of memorable times and places. However, unlike tea-towels and spoons, snowdomes go far beyond being just a mere souvenir. They combine romanticism, reality, and illusion to conjure up miniature utopian worlds. Through the distorted glass of the snowdome the Australian landscape, and all that it symbolises, can be portrayed in a myriad of optimistic and ironic ways.

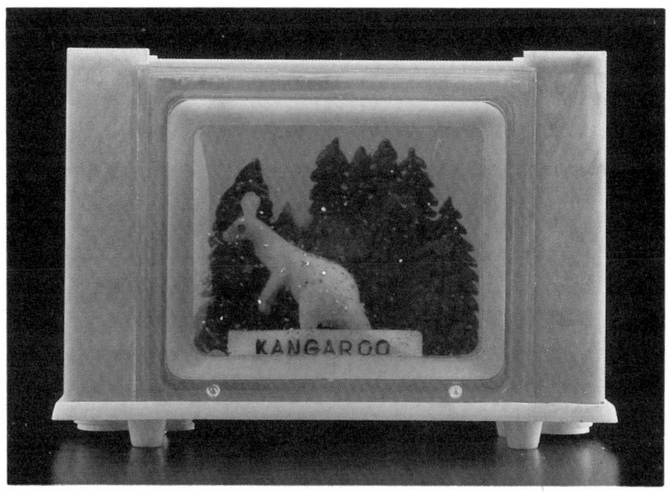

Kangaroo **The more bizarre the snowdome, the greater its appeal seems to be Exhibit no. 126**

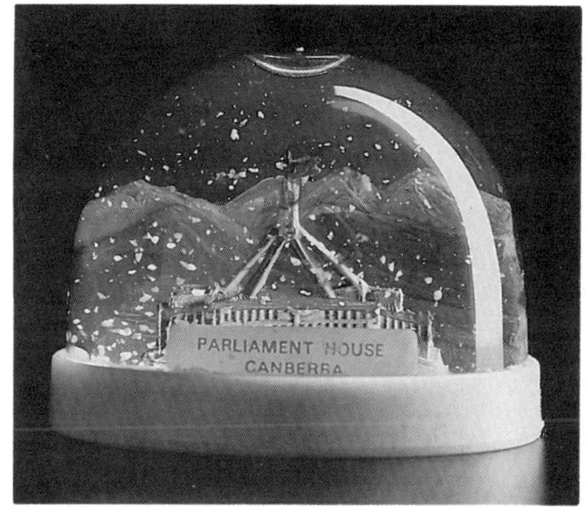

Parliament House, Canberra
Its ambiguous flag reputedly makes it a collectible Exhibit no. 110

THE ART OF THE TRAVEL POSTER — Katie Spearritt & Anthony Hayes

In mid -1990 Australian Airlines launched a series of 12 travel posters which aimed to mould a deliberately up-market image of the nation. The new series, according to airline executives, attempted to depart from the 'ordinary tourism poster' — the traditional shot of a scenic spot. Although the designer knew some images were daring, he was unsuspecting of the censorial storm that followed.

A Gold Coast promotional poster showed two semi-naked women and a man in bed on the balcony of their high-rise apartment overlooking Surfers Paradise. Not only was it a more sophisticated focus on the tourist, rather than destination, but it gave a new twist to the 'sun, sex, surf' genre that has unashamedly sold Queensland's Gold Coast for decades.

It was not received favourably by passengers, airline staff, or the public. Design aesthetics and ethics of the travel poster captured media attention across the nation as never before. Australian Airlines soon withdrew the poster.

The publicity and its subsequent withdrawal could have been an advertising masterstroke for the airline. The poster was not aimed to sell itself, however, but to sell a new image of the Gold Coast. The old glitz had seemingly gone, and with it the hottest ticket in town — a Broadbeach Hotel revue of topless showgirls. That image was first forged in the 1960s when meter maids and pyjama parties accompanied large-scale hotel development and neon extravaganza.

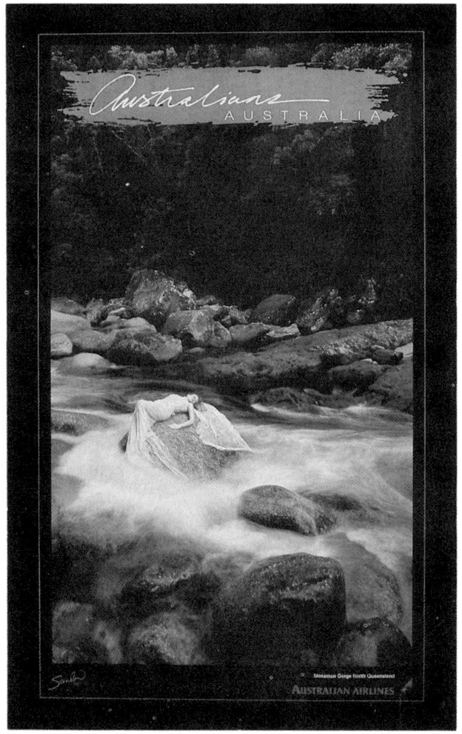

***Mossman Gorge*, 1990 A dramatic contrast to the traditional tourism poster Copyright: Australian Airlines Exhibit no. 46**

The Gold Coast of the 1990s has borrowed some ideas from image-obsessed Las Vegas. Low-rise resorts, theme parks, casinos, international hotels, and a proudly elitist set of 'Sovereign Islands' make this 42-kilometre golden strip a playground for Australia's sophisticated holiday makers. Australian Airlines was set to capture this market, but the market was not ready to accept this image.

Inventing new ways to characterise the continent and its people has always been problematical. The Gold Coast poster episode revealed the ways in which travel posters reflect how Australians view themselves and their land. Rather than direct this perception, the attempt pushed the boundaries too far, too soon.

Although Australian poster artists have traditionally had a free hand in constructing the sight or sights depicted, the site is usually chosen by clients such as airlines, tourist commissions, and travel bureaus, who have interpreted their role as selling a specific destination. Consequently, few travel poster artists have ventured beyond the safe imagery of mainstream landscape icons. The portrait of the beach, the outback, or the bush is instantly recognisable and offers beguiling delights of leisure and pleasure. Not surprisingly, the poster's ambit of representation has rarely stretched to critical reflection of the diversity of Australia's population.

National stereotypes have featured in travel posters since the 1930s, when the poster first emerged as a powerful advertising medium. Australian industry and government leaders articulated the notion of travel as economically beneficial in 1928 with a *Proposal to develop the Tourist Business of the Commonwealth and to stimulate greater travel amongst Australians*. The report focused on the selling of the landscape to foreign investors:

Travel is a commodity that requires selling, more particularly in the case of a young country like Australia which, in a comparative sense, is little known to the large masses of people overseas; then, too, our remoteness from the great centres of population in the English speaking world demands greater zest in salesmanship if we are able to obtain that share of the world's tourist business which our unique resources and scenic beauty merit.

Australia was perceived as the 'Cinderella of the world's tourist business' by the Australian National Travel Association (ANTA) promoters.

Many of the posters produced by ANTA in the 1930s were colour lithographs or photolithographs, unlike the modern photographic poster. Surfing beaches, fauna and flora, mountain resorts, and the Australian outback featured as distinctive icons of Australia in assorted travel literature.

Over the next two decades the travel poster continued to develop, both as an art form and as a means of promoting Australia as a holiday destination. However, the intrinsic 'Australian' character of the posters, as depicted by the concentration on the landscape itself, did not radically alter.

DESIGNER POSTERS

In the 1980s travel posters reflected greater graphic adventurousness with the rise of the graphic designer. Popular artists such as Ken Done, Anne Graham, Martin Sharp, and Jo Anne Hook moved toward a more visually stimulating presentation of the land-

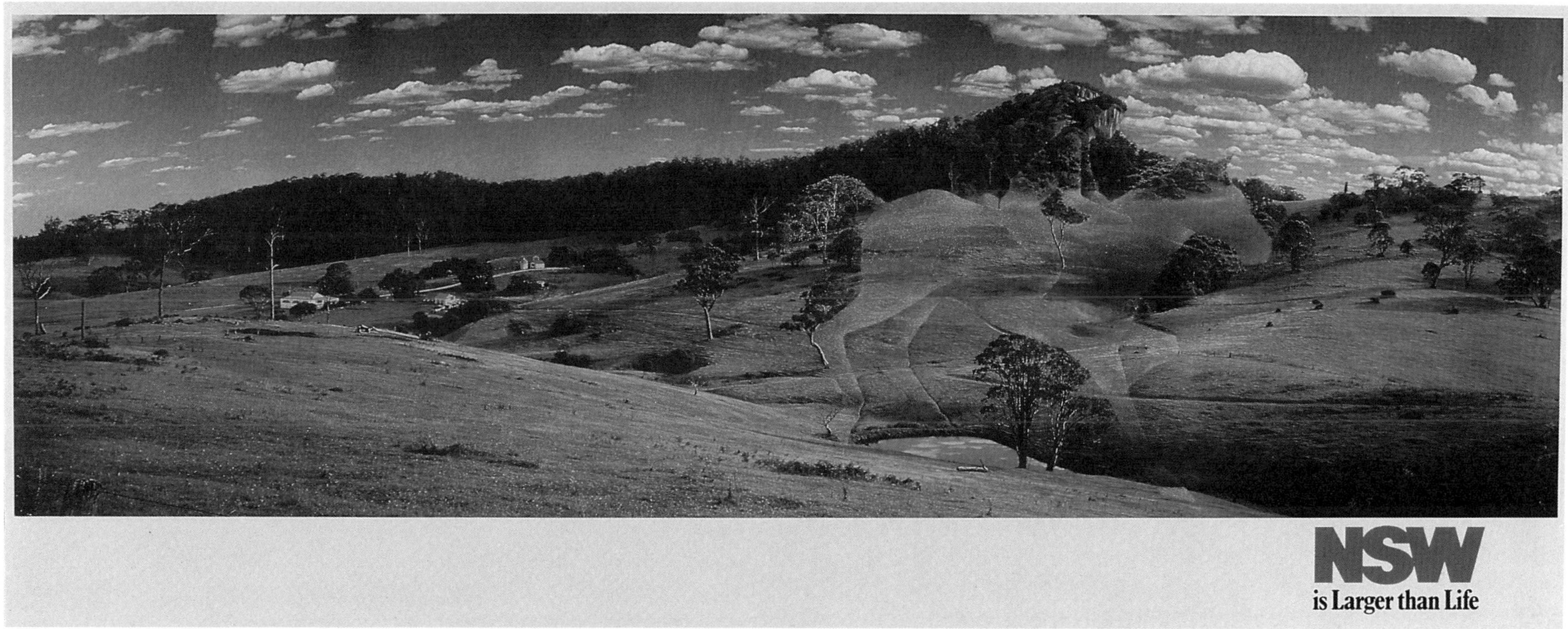

New South Wales is Larger than Life **The Australian landscape assuming a surreal perspective Copyright: NSW Tourism Commission Exhibit no. 37a/b**

scape, without radically reconstructing perceptions of the landscape.

Travel posters became less visibly 'industry' oriented during this period. Stylish artistic posters replaced the 'cute koala' print campaigns of the early 1980s. Thus Anne Graham's flagship TAA series in 1985 was designed as much for bedroom walls as for the travel agent's offices.

This recognition gave the process of marketing Australia a new sense of itself. Posters created by the New South Wales Tourist Commission, for example, created a surreal effect by superimposing images on the landscape. And Jo Anne Hook's 1988-90 series offered a touch of the tropical exotica by interweaving Australia's flora and fauna in unusual patterns.

As in earlier eras, the photographic travel poster squarely rested on the presentation of a comfortable brand of Australianness centred on pastoral, outback, and coastal images.

The travel poster's emphasis on the natural landscape in the 1980s had market-driven imperatives. Heightened environmental consciousness among the public was best captured by a three-poster series advertising Canberra as 'Australia's natural capital', replete with blue gums and wattle.

Added to this, Australia's natural sights have been identified by recent Australian tourism research as the most effective visual icons to attract international visitors. Tourists may make the final decision to visit Australia based on their desire to watch or take part in an event, or for a variety of other specific reasons, but the prime motivation for travelling to Australia is to experience the landscape.

Clearly, artificial 'themes' like the Australian Tourism Commission's Australian Year of Sport in 1993 are unlikely to surpass the landscape as the dominant influence of travel advertising. Yet the 'unique resources and scenic beauty' described by industry pioneers are threatened as the century closes. The challenge that now confronts the travel industry is to find new and acceptable ways of discarding our 'Cinderella of world tourism' image.

HANDBACKS & PHOTOJOURNALISM — Ann Stephen

Newspaper photographs construct events as history and produce many of the images that enter popular memory. Photography's claims for truth — 'the camera doesn't lie' — and value, 'a picture's worth a thousand words', make it a particularly powerful form in the mass media.

In Australian photojournalism, landscape imagery appears most often as 'rural crisis', 'environmental struggle', or land rights. It is particularly in the representation of the latter that photography has played so prominent a part and occupies such an ambivalent relationship to its subject.

The mediated nature of photographic 'truth' is usually well known to the producers and their subjects, but is lost when a single image comes to stand for a complex history. Two mass media photographs commemorating major handbacks of land to Aboriginal communities deserve closer scrutiny in this regard.

The first, taken by Mervyn Bishop in 1975, is shown on the opposite page. Gough Whitlam, in buttoned suit, leans towards the blackfellow, consciously enacting an historic moment. The pouring of soil from hand to hand is a highly symbolic act, for old men do not usually play with sand. The photographer has crouched down low in front of his subjects creating a monumental group.

Vincent Lingiari, representing the traditional land owners, standing in overlarge shirt and pants, is more inscrutable. We are unused to his expressions, and he is less used to all

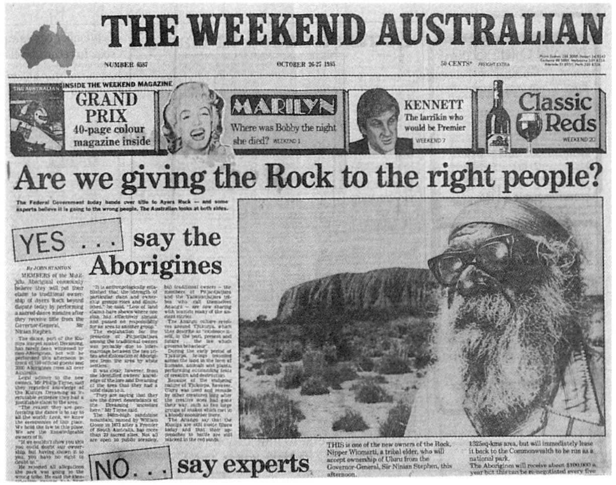
THE WEEKEND AUSTRALIAN

NUMBER 6587 OCTOBER 26-27 1985 50 CENTS*

INSIDE THE WEEKEND MAGAZINE

GRAND PRIX 40-page colour magazine inside

MARILYN Where was Bobby the night she died? WEEKEND 1

KENNETT The larrikin who would be Premier WEEKEND 7

Classic Reds WEEKEND 20

Are we giving the Rock to the right people?

The Federal Government today hands over title to Ayers Rock — and some experts believe it is going to the wrong people. The Australian looks at both sides.

YES ... say the Aborigines

NO... say experts

THIS is one of the new owners of the Rock, Nipper Winmarti, a tribal elder, who will accept ownership of Uluru from the Governor-General, Sir Ninian Stephen, this afternoon.

Are we giving the Rock to the right people? Weekend Australian, 1985 Copyright: *Weekend Australian*

this attention. We cannot read him the way we know Gough. Both heads, framed against a clear bright sky, direct our gaze down to their hands. Beneath Gough's white clenched fist a small almost geometric pile of dirt has formed in the outstretched black palm. The biblical references resonate in this act, of dust to dust, of Gough appearing to play God in a twentieth-century version of Michelangelo's *Birth of Adam*.

On the twenty-fifth anniversary of the ceremony, Whitlam recalled that:

I bent down and picked up the earth and I poured that earth into your leader's hand. On behalf of all Australians I handed the earth, the soil, the land back to the Gurindji.

This act of handing back the soil was of great significance to both men. Eleven years earlier, Lingiari led a strike by hundreds of Aboriginal stockmen and their families against Lord Vestey, their English boss who was then the largest land owner in Australia. This legendary walk-off to protest at appalling living and working conditions rapidly accelerated into demands for the return of the Gurindji's traditional lands.

'The 1970s was an era of work for Aboriginal people. Things had to be done, and fast — there were organisations to establish and land to get back. See Prime Minister Gough Whitlam — the air literally crackled', Tracy Moffatt commented on the photograph in Mervyn Bishop's exhibition catalogue, *in dreams*.

In August 1975 the Whitlam government bought, for the traditional owners, part of Wave Hill station which has since reverted to Aboriginal freehold by way of a land claim. While Whitlam had not forgotten the gesture, the photo was in fact a re-enactment organised by photographer Mervyn Bishop. The handback ceremony had just been performed in the dark and unphotogenic bower shed, seen in the right of the photograph. Bishop, who was staff photographer for the Department of Aboriginal Affairs (DAA), asked the two leaders to stand outside against the blue sky and pour the red sand from hand to hand again. 'I was taking a sort of public relations picture not unlike the *Herald* press photos which were used in DAA's magazine', he said. Perhaps because it was a repeated act, Gough's gesture was more than usually meaningful. To Bishop's frustration, *Aboriginal News* reproduced this dazzling colour photograph in black and white. However, its economy as a symbol of reconciliation has made it one of the most reproduced images of the Whitlam era and of the Aboriginal land rights movement.

FROM AYERS ROCK TO ULURU

A decade later, another great milestone in the Aboriginal movement was the handback of Ayers Rock to the traditional owners, with it reverting to its pre-contact name, Uluru. The decision was opposed by an extraordinary publicity campaign led by the Northern Territory Government and picked up by Rupert Murdoch's national newspaper, the *Australian*. While purporting to express an impartial position, the debate over ownership was expressed in terms of 'white experts' versus Aboriginal people (implicitly not experts).

On 26 October 1985, the day of the handback, the *Weekend Australian's* front page was dominated by the headline 'Are we giving the Rock to the right people?'. This appeared above a photograph representing Uluru in the standard postcard format, except for one major difference. Almost half of its right side was obscured by the head and shoulders of an old Aboriginal man. He was shot at very close range, implying a sense of threat, suggesting he was 'too close for comfort'. The fact that he blocked off the Rock visually reinforced the fear that Aboriginals were taking it away from 'all Australians'. Furthermore, the caption names him as 'one of the new (note the word 'new' not traditional) owners of the Rock, Nipper Winmarti, a tribal elder'. He is photographed in Western shirt and glasses, confirming the views of those 'experts' who claimed these were not 'authentic Aborigines' and the new arrangement was a 'European-style legislative organism, which steps outside Aboriginal tradition'.

The Rock has a very recent history as a white 'sacred site'. In 1920 it was part of the land gazetted as an Aboriginal reserve. Only in the 1950s did commercial interests begin a concerted campaign to exploit its tourist potential. By 1951 Connellan Airways had applied to build an airstrip to land 'for an hour or so with people who wanted to photograph the Rock', and in 1958 Uluru was excluded from the Reserve for tourism. It was this decision which was to transform the lives of local Aboriginal communities.

With the Red Centre booming as a tourist destination in the 1960s, the Assistant Director of the Welfare Board felt compelled to write to bus operators asking them to stop their customers paying Aboriginal people for artefacts or taking photographs, commenting that 'the few shillings Aborigines make in this way tends to undo the work of my officers. Giving money to people has an adverse effect in our attempt to teach the natives the principle of work for pay and makes them parasites on society'.

In 1971 Paddy Uluru, one of the Yankuntjatjara elders, spoke to government officers about the desecration of sacred places at Uluru by tourists. This precipitated a laborious process through which local Aboriginal communities eventually were able to prove their knowledge of the land, and later formed the basis of their successful land claim.

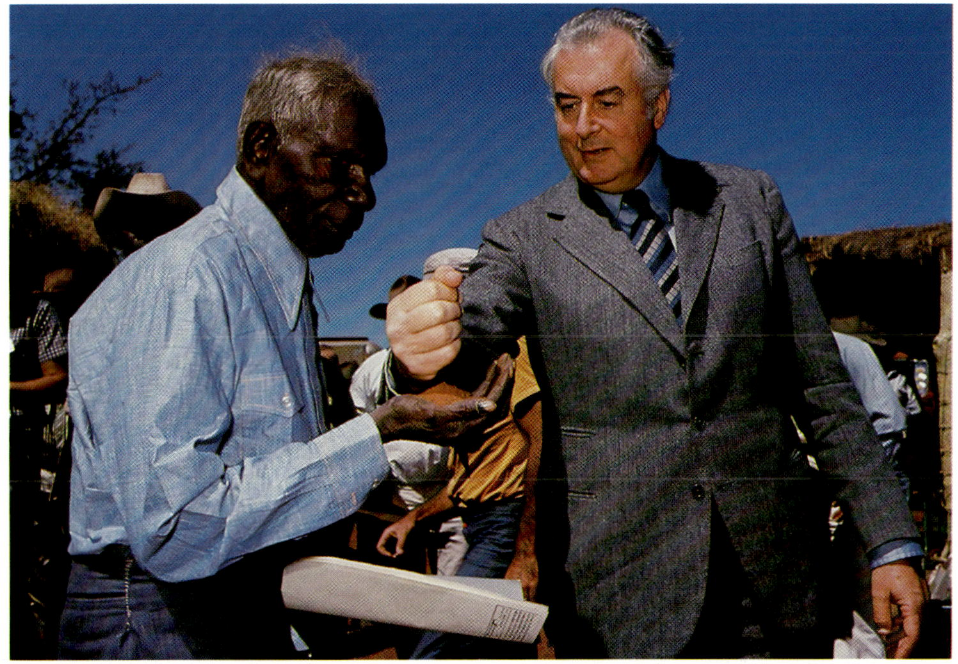

Prime Minister Gough Whitlam pours soil into the hands of traditional landowner Vincent Lingiari, NT, **1975 Copyright: Mervyn Bishop**

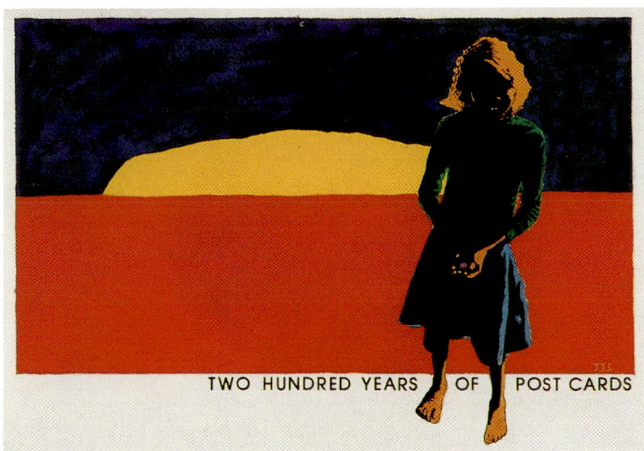

***200 Years of Post Cards*, designed by Paul Worstead**

Fears the *Australian* sought to foster, that the Rock was being 'given to the wrong people' and would become inaccessible to white tourists, have proved groundless. Tourist photography, which had been one of the main incursions into the sacred sites, has become regulated. When buses now stop at the Rock, they are confronted by the following sign:

Photographs: Traditional owners request that you respect their customs and refrain from photographing local Aboriginal people. Having one's photograph taken is considered culturally inappropriate as the image captured on film, believed to be part of one's spirit, is removed from that person's control forever. Thank you for your cooperation.

It has been suggested by photographic historian Geoffrey Batchen that this may be a strategy adopted by the Aboriginal community at Uluru, to turn 'the discourse of primitivism against itself to secure the privacy that most Australians would assume to be their automatic right'. Certainly, the return of land has secured a space for Aboriginal people to consider a range of issues, among them tourism and photography. Having been subject to the European stare through the camera lens for more than 100 years — first in the name of ethnography, later as tourism — it is appropriate that Aboriginal people are now able to exercise some control over their representation.

IMAGINING UTOPIA: THE SELLING OF SUBURBIA — Kimberley Webber

Australians continue to define themselves in terms of the bush, yet their common experience is that of the suburb. And the suburb is both their greatest aspiration and their worst nightmare. For most of this century, it has been the popular butt of comedians and writers seeking to distance themselves from their birthplace. In the guise of Edna Everage, Barry Humphries (born the son of a builder and property developer in the middle class Melbourne suburb of Camberwell) is our most vicious and most accurate observer.

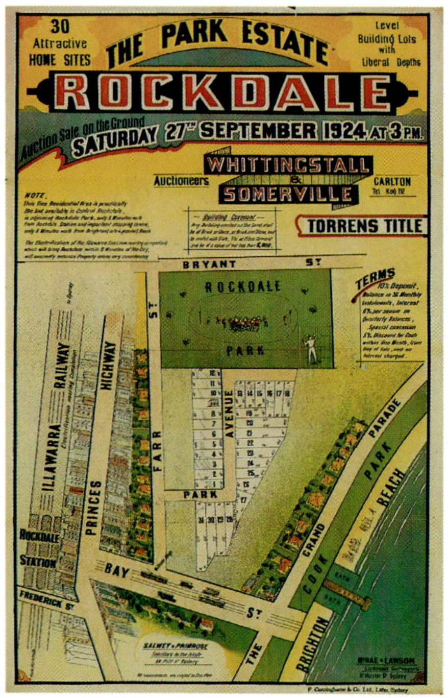

***The Park Estate, Rockdale*, 1924 A rural Arcadia with ready access to the metropolis**
Exhibit no. 28

This paper focuses on two images from the exhibition which sell the ideal of the suburb as 'bourgeois utopia', an escape from the horrors of dirt, overcrowding, pollution, and noise which have come to characterise the city. These images are The Park Estate at Rockdale (1924) and New Choices in Housing — The Green Street Joint Venture (c. 1990).

Rockdale lies on the shores of Botany Bay. Originally called West Botany, its name was changed in the 1870s by the local postmistress, Mary Ann Geeves. The Illawarra line has provided a direct rail link with the city since 1884. Reporters on the first train from Redfern described the then infant suburb in glowing terms:

Those who spend their leisure time in this Arcadia will reap many pleasures and advantages from its surroundings and resources.

But the Arcadia viewed by these journalists was not the same as that seen by Joseph Banks. In 1770, he had lyricised over the area's woods, fresh streams, 'wholesome air', and abundance of birds and animals. A century later the reporters observed that such 'old, wild and picturesque scenes' were rapidly disappearing, replaced by a suburban Arcadia of 'neat homelike cottages' set among orchards and gardens filled with European trees and flowers.

By 1921 the infant suburb had grown to a municipality of 8585 dwellings with 39,935 residents. In the ten years to 1931, the population and the number of dwellings almost doubled, as Rockdale, like the rest of Sydney, experienced a speed and intensity of development not seen again until the 1950s.

A reading of the Park Estate poster says much about the character of this development and the nature of suburbia itself. The Park Estate was to be auctioned on Saturday 27 September 1924. Spring was the season for land sales, as the *Evening News* reported on 25 August 1920:

Now that the weather is warming up a resumption of sub-division sales is imminent . . . The demand is here already. Whatever may be said of the improvidence of the Australian, he nonetheless has sufficient commonsense to acquire his own home or at least strive hard for it.

In all probability the estate was formed from the grounds of one of the large 'villa residences' built in Rockdale in the nineteenth century. However, the Park Estate is rather a grand title for what was little more than a street block with only one of the 30 blocks in the subdivision being directly adjacent to the park. But the name does suggest the selling power of the garden suburb ideal in post World War I Australia.

Significantly, in a promotional map with few details, the only buildings pictured are the freestanding residences with established gardens along the Grand Parade and more modest cottages, again with trees, on Farr Street. Although blocks are comparatively small (all are 40 foot wide and length varies between 70 and 277 feet), purchasers are required to erect a certain standard and size of dwelling. No timber or fibro homes, and no tin roofs, are to be built lest they lower the 'tone'.

But this was no rural retreat from the city. While the auctioneers allow some romance to creep into the title, they are quick to point out the close links between Rockdale and the city. With only a five-minute walk from Rockdale station, once electrification of the railway is complete, the auctioneers claim that the fortunate owners of these allotments will be just 12 minutes from the city. The accompanying text describes these blocks as 'practically the last in central Rockdale'.

The means of getting out of the suburb dominate the auction poster, both visually and textually, more than any of its 'natural' and human-made resources. Indeed, no reference is made to schools, community centres, or other local facilities, and Brighton Beach is viewed with some caution. 'Safe and popular' is the best that can be said of it. The purchasers of these blocks are not seeking a rural Arcadia, but ready access to the metropolis. Indeed, 'nature' makes them nervous.

A further characteristic of urban development in the 1920s was the appeal of making money. Land speculation was rife in Sydney and, with electrification of the railway due for completion in 1926, part of Rockdale's attraction was the likelihood that its property values would rise significantly.

NEW CHOICES IN HOUSING

Seventy years (and three property crashes) later, New Choices in Housing — The Green Street Joint Venture carefully avoids any suggestions of profit or land speculation. This project is a joint federal, state, and local government venture which aims to promote better and more cost-effective use of land in Australia's suburbs. A number of residential developments have been sponsored in major cities to test ideas about size, arrangement, and siting of houses, planning of streets and roads, development of open space and vegetation. In 1983 the National Capital Development Commission worked with the Joint Venture to develop two demonstration estates in Canberra: Isabella Plains and Calwell. The poster opposite illustrates a typical street in one of these estates.

Like the Park Estate, New Choices in Housing is a somewhat grand title. Indeed, these apparent 'new choices' look exactly like any other suburban street in Canberra. Devoid of boundary or other distinguishing features, the front gardens blend blandly into one another. The dominant features of the photograph are not new homes, but letterboxes, driveways, and the road. The viewer could suspect that so little attention has been paid to the actual houses because individual details do not seem necessary when they all look so alike.

However, a reading of the Green Street literature makes it apparent that this particular composition was taken to illustrate what are seen as innovations in suburban design. The lots are smaller and open space is maximised by placing the houses on the edge of the allotment. All have provision for off-street parking. The lack of demarcation between houses and the road reserve is intended to 'encourage landscaping by residents and . . . engender a sense of ownership of the public spaces'. Away from major through-streets, roadways are narrow to slow down traffic and make roads safer for children and pedestrians. The latter is particularly important, since there are no pedestrian pathways. All these innovations make houses cheaper to build, more cost-efficient to run, and concentrate more in less space.

They also reinforce the very 'suburban' qualities that writers like Robyn Boyd have so railed against. In 1952 he wrote in *Australia's Home* that:

The Australian town-dweller spent a century in the acquisition of his toy: an emasculated garden, a five-roomed cottage of his very own, different from its neighbours by a minor contortion of window or porch — its difference significant to no one but himself.

Forty years later the 'toys' have become larger and more energy efficient but are still no less emasculated or indistinguishable from each other than they were in the 1950s.

Characteristic as much of Canberra as of this particular suburban development project is the background of the bush — a mixture of pines and eucalypts — at the end of the street. Aiming to create a city 'unlike any other in the world', post World War II development in Canberra was implemented around the National Capital Open Space System. Suburb and open land were to be closely integrated through a series of satellite towns.

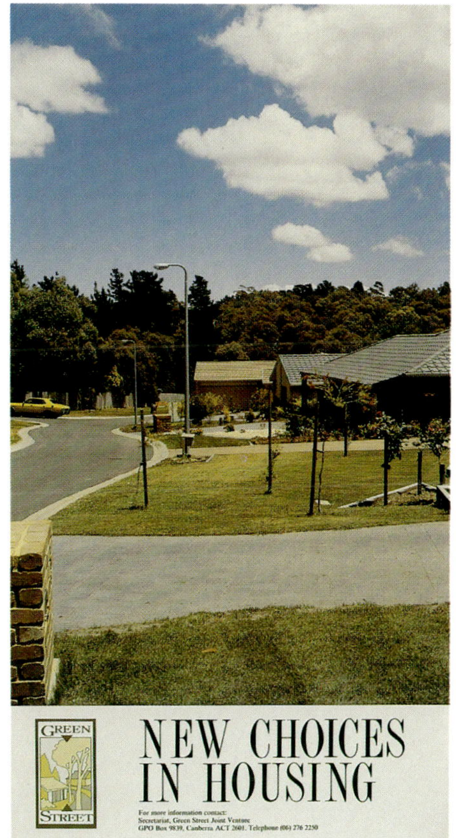

***New Choices in Housing*, 1990 Suburbia, a bourgeois utopia? Exhibit no. 30**

In *Canberra, From Limestone Plains to Garden City* (1986), the National Capital Development Commission claimed:

There are few cities anywhere, planned or otherwise, that fit into their natural environment as well as Canberra does. Urban development is subordinated to topography; respect is paid to trees and vegetation, to the greening of the city and open spaces of great variety are ubiquitous.

Suburbs are isolated from offices, factories, hospitals, and shopping malls by freeways or tracts of pine forests and bushland. A crucial factor in living in such suburbs is therefore the ability to get about. And while Canberra does have an extensive bus system, the car remains the easiest means of getting out. As a consequence, Canberra has become not so much the 'garden city', as envisioned by Walter Burley Griffin, but the car city.

These images sell the idea of suburbia as a 'bourgeois utopia', offering both refuge and domestic bliss. Yet indicative of the ambivalence we have to suburbia is the emphasis given within each poster to the means of escape from it, away from what Edna Everage has described so movingly as her 'favourite things' (sung to the tune made famous by Julie Andrews in *The Sound of Music*):

White cabbage-moths round tomato plants flitting / The swish of a sprinkler, the smell of new knitting / Tea leaves on the maiden-hair, silverfish in the bath / The sugar ants' trail up my front cement path . . .
These are a few of my favourite things.

Lest we succumb to the siren song, the Kingswood must be kept ever at the ready, and road, rail, and major highways be always close at hand.

THE LANDSCAPE QUESTION — Ian Burn

An opal-hearted country,
A wilful lavish land —
All you who have not loved her
You will not understand —
Though earth holds many splendours,
Wherever I may die,
I know to what brown country
My homing thoughts will fly.
Dorothea Mackellar, *My Country*, 1908

The land holds a distinctive place in the cultural experience of the majority of Australians, black and white. The Aboriginal peoples, with more than 40,000 years of attachment to the Australian environment, have developed a relationship with the land which remains central to their social, religious, and cultural life. Aboriginal imagery manifests a sense of custodianship of the land and records complex stories of journeys, mythical origins, sacred sites, and daily incidents scattered across an endless landscape.

White Australians claim two centuries of rapprochement with the land, but the complexity of the process is exposed by the need for constant renewal — of artists reconstructing their practices, even themselves, in relation to the landscape. Before European discovery of Australia, the logic of a great southern landmass fed the imaginations of northern cultures for centuries. Long after white settlement, the *idea* of Australia retained a fictional life which far outstripped any scientific rationale.

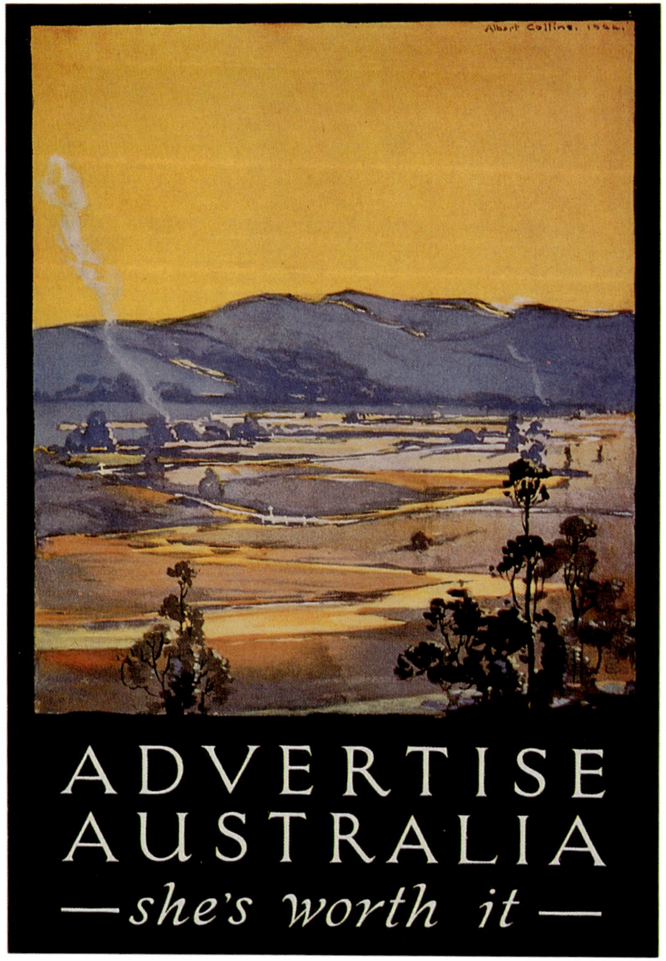

***Advertise Australia*, 1924 Designed by Albert Collins**
Australasian Advertisers' Manual & Newspaper Directory

Many believe it still does. The process of mapping the landscape into the imagination continues to create visual metaphors for the new society, a metaphysics for signifying difference. In a country urbanised before the landscape was colonised, a 'geography' of the land continues to be shaped as much by imagination as by reality.

Pictures of the landscape are tangible objects specific to place, objects which can be possessed and which symbolise a cultural possession of the land. What, then, do artists in Australia paint when they paint 'the landscape'? Over the past two centuries the landscape has been invented and reinvented many times, in the context of serving different cultural, social, and political needs. The *bush* had a cultural meaning in this country prior to the *landscape* acquiring a specific form. As cultural concepts, both terms exist as fictions of convenience, yet much artistic endeavour has focused on attempts to define the bush in terms of a concept of the landscape.

References to the bush were taken for granted as early as 1803, while an English outlook pervaded; but in the new environment its meaning took on other significances, referring generally to unsettled or sparsely settled areas, a kind of 'anti-city'. It is not hard to understand that the relationship between Sydney and the bush in the early nineteenth century was far less secure and implied greater threats than the relationship between town and country in Britain.

By the end of the nineteenth century the idea of the bush had become encased within concepts of the landscape — naturalistic conventions had been adopted and adapted to render the bush pictorially. Expressing a confident, optimistic, white presence, the painters — to use 1920s terminology — had created a national sentiment for our landscape, its forms charged with a national significance. The process, however, presupposed an emotional bonding of people and place; an understanding of this 'wilful lavish land' would still be denied to those 'who have not loved her'.

Since the late nineteenth century, the Australian landscape has been held up as a mirror for a 'national psyche' far more than as a mirror of the 'human soul'. In the dominant white culture this has granted the landscape a value perhaps beyond that of any other country, making it a potent obsession within the intellectual culture of Australia. Yet strip away its visionary or fictional character, and what is uncovered is a psychological need in relation to the *idea* of Australia, not to some ultimate 'real' nature or 'authentic' Australia.

BHP & THE VILLAGE PEOPLE — Charles Pickett

Corporate reports are no-expense-spared productions. The 1988 *Report to Shareholders* of Broken Hill Proprietary Co. Ltd (BHP), for example, boasts photography by David Moore and design by Emery Vincent Associates, two of the leading graphic taste-makers in Australia today.

Not surprisingly, BHP's report contains some striking photographs, including the one reproduced on the next page. It is captioned in rather matter-of-fact terms: 'Stretching into the distance, cranes work to complete Australia's largest resource development, the North West Shelf Project. At maximum production it will ship almost six million tonnes of liquefied natural gas (LNG) a year.'

There is presumably no need to elaborate on the masculine automaton striding towards the camera, bestriding the spreading refinery (itself bestriding land and sea) like an industrial colossus. He and the factory stand at the end of a graphic tradition spanning a century or so. Yet this imagery holds such a secluded place in our culture that most viewers would place the automaton (with his sunglasses and unbuttoned, sleeveless shirt) with US disco band the Village People (or Oxford Street venues such as Midnight Shift or the Tool Shed) rather than BHP. The gay stereotype has overwhelmed the heroic icon.

Celebratory images of industrial buildings, technology, and workers are largely confined to the pages of shareholders' reports, such as those issued by BHP. Confined and secluded, but hardly marginal in economic and political terms.

Such images once enjoyed a more public place. During the early decades of Australian manufacturing, industrialists routinely employed boastful images of factories in their advertising. At the same time, unionists seized on the imagery of heavy masculine labour for their banners and posters. Apart from this period of visibility, factories and mines have seldom fitted comfortably into the iconography of Australian landscape and townscape. Some writers have ascribed this to Australia's 'rural consciousness', but ambivalent reactions to industry are not unique to Australia.

When, in the early 1800s, cities first appeared with skylines dominated by chimneys and warehouses rather than church steeples and town hall clock-towers, writers and artists came from far and wide to see them. It was immediately recognised that places like Manchester and Lowell embodied a new type of city and civilisation. It was also recognised that these cities displayed a fundamental contradiction in the close coexistence of unprecedented wealth with abject poverty, not to mention environmental degradation.

As a result, Marx and Engels were far from the only English 'opinion-formers' to encourage an ambivalent attitude to industry. Even English captains of industry took on the trappings of landed gentry as soon as they were able, while in the twentieth century the same ambivalence inspired the flight of the masses to non-industrial 'garden' suburbs.

The Dickens-reading anglophiles who ruled Australia invested heavily in the creation of a non-industrial rural utopia, gaining their wealth from feeding and clothing the industrial masses of the old world. Both this archetype and the alternative vision promoted by manufacturers and unionists receded before the United States' inspiration of consumer-orientated industrialism, which fetishised products rather than production.

THE INVISIBLE MINER

Today, Australian economic culture contains a strange irony: while manufacturing success is widely accepted as the measure of national achievement — witness fears that Australia is becoming a nation of waiters on United States and Japanese tourists — industry and industrial work are all but invisible in the cultural mainstream, if not in the corridors of power. The culture of the producers has retreated to industrial towns and company reports, and not solely because of the triumph of consumerism. The strong, masculine, manual imagery of the labourist view failed to adapt to the changing nature of manufacturing work and workforces. BHP's automaton, not coincidentally a construction worker rather than a factory operative, represents one of the few surviving conformists to the labourist archetype.

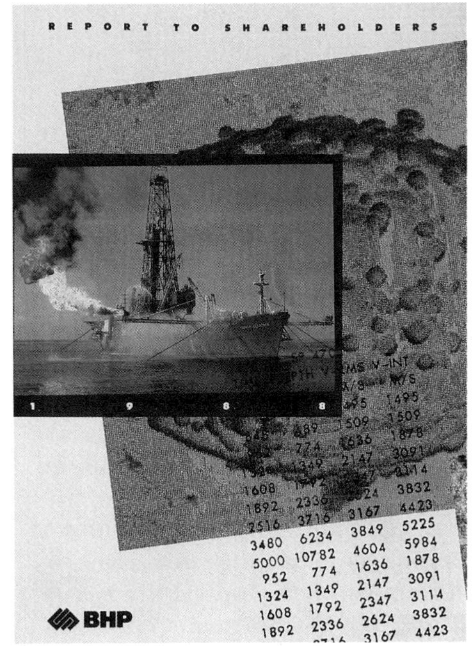

***BHP Report to Shareholders*, 1988**
Copyright: BHP

(The Village People made the same judgment, so BHP is on safe ground.) Similarly, BHP's refinery is more emblematic of the company's activities than the frequently rented (rather than purpose-built), and hence non-descript, factories of today.

Although capable of maintaining the graphic archetypes, BHP's representation of industry has undergone telling modifications. In its corporate publications, workers appear with increasing frequency, and they seldom appear as automatons. Instead, they are given names — often female or indigenous names — in marked contrast to the one-dimensional boardroom portraits. No doubt, the design and advertising companies which produce company reports insist on the need for a 'human face' and an appearance of environmental responsibility. Indeed, rainforests and mountains outnumber factories in some corporate publications.

Attempts to give industry a rosy glow are related to another contemporary development. There are signs that industrial imagery may be about to gain a new lease of life. The economic and technological obsolescence of so much Australian manufacturing has created acres of redundant industrial sites. In both the United Kingdom and United States this process has produced a new appreciation of the heritage and history of industry, whereas Australian conservationism continues to be dominated by anti-industrial rhetoric. Although increasingly anxious to preserve and profit from their domestic architectural heritage, Australians have shown little inclination to value industrial architecture.

Nonetheless, the campaign to save Sydney's Finger Wharf has revealed a new appreciation of the industrial aesthetic. Future developments in conservationists' attitudes are difficult to predict, given the social structure of the movement. When the industrial heritage movement began to boom in the United Kingdom, the novelist Angela Carter noted that industrial townscapes were:

seductive principally to the bourgeois romantic intellectual who sees them with the fresh eye of one not born and bred in a back-to-back house . . . On the other hand, the history of taste may well be that of the obscure and probably warped predilections of the bourgeois romantic intellectual . . . After all, only a handful of eccentrics enjoyed mountains until a mountain got up and followed Wordsworth across a lake. ('Industry as Artwork', *Nothing Sacred: Selected Writings*, Virago, 1982)

Given that a lot of people have been enjoying mountains lately — even BHP, if its annual reports are to be believed — it may be that factories will win a place in the heritage canon. After all, ex-factories no longer produce poverty, exploitation, and muck. As heritage, factories can join BHP (and the Village People) in celebrating industry.

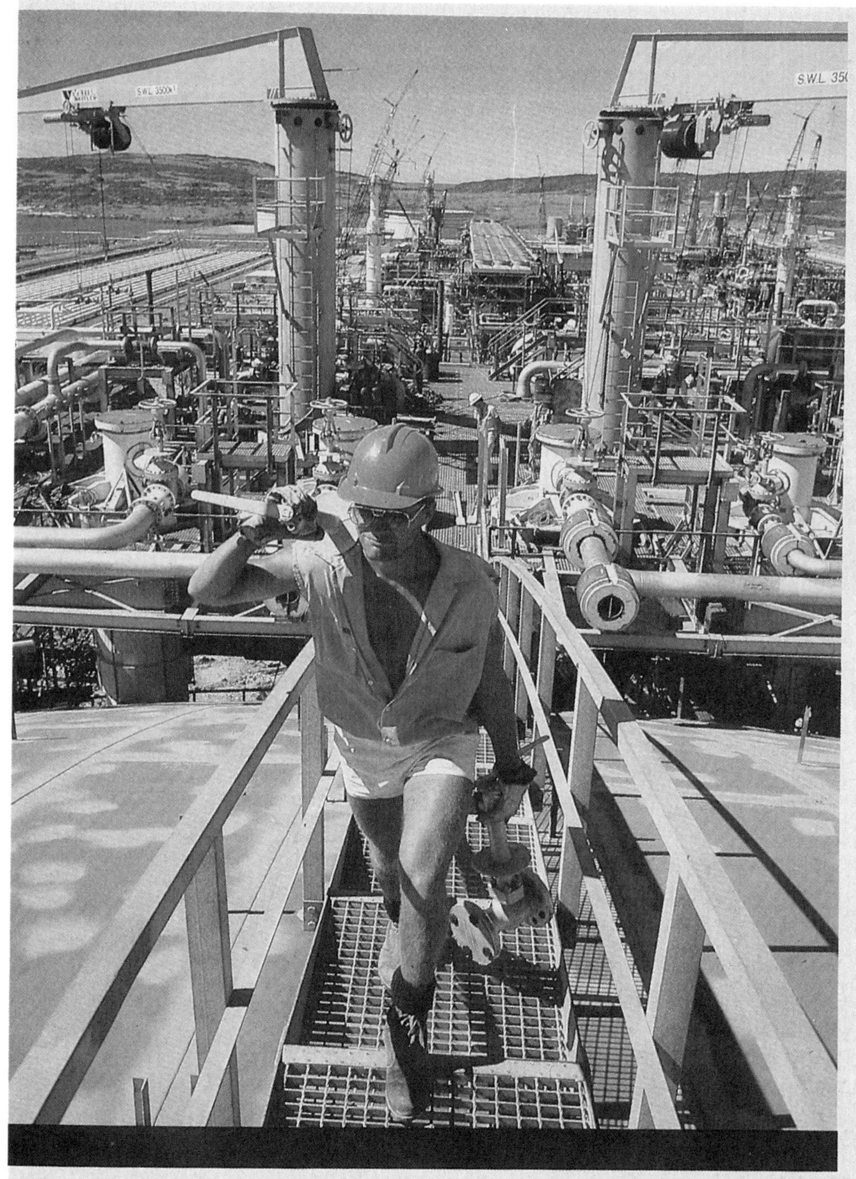

Gay stereotype or heroic icon?
'Masculine automaton striding towards the camera, bestriding the spreading refinery' ➤
***BHP Report to Shareholders*, 1988**
Copyright: BHP

Quarterly profit results ($000) Quarter ended

THE ABORIGINAL FLAG AS A SYMBOLIC LANDSCAPE — Sherry McLaughlin, John Bernard Tiparui, Terry Sams

THE ABORIGINAL FLAG (1)

The Aboriginal flag was designed by a man named Harold Thomas. Harold Thomas was born in Alice Springs, but grew up in white society in Adelaide, where he became involved in Aboriginal politics. He designed the flag in 1971.

Harold believed that the flag would unite the whole of Aboriginal Australia and give Aboriginal people a sense of belonging. The black on the flag reminds you that Australia was once all black. The yellow represents the sun — so powerful — giving life to everything on earth. The red symbolises red ochre and the colour of the country.

Harold Thomas feels very proud of his design. He knows there are certain Aboriginal family groups in the centre of Australia, who do not accept this flag. They already have their own symbols and totems and Harold respects that.

As I was brought up in a white society and do not know my family's symbols or totems, I think that the flag is a great thing. I can look at it and say 'I am part of that, I am Aboriginal'.

I think that the most important function of the flag is the fact that it unites the whole of Aboriginal Australia.

Sherry McLaughlin, Geraldton WA *

THE ABORIGINAL FLAG (2)

A long time ago the Aboriginal people in Australia did not have a flag. Harold Thomas made the flag because there was no sign to represent all Aboriginal people. Harold put a lot of thought and planning into the flag before he designed it. The colour and shapes on the flag mean red for land, black for people, and yellow for the sun.

Although I am Tiwi with a different culture to that of mainland Aborigines, I still feel at one with all Aboriginal people. Our community feels that the flag is good. Harold Thomas felt that he brought Aboriginals together in the flag. It is good to have a flag as it represents our own country, and it is very important to us all to have the flag and we should feel proud.

I think the flag gives all Aboriginal people a feeling of being stronger together — that we are one people and proud of ourselves.

John Bernard Tiparui, Nguiu *

THE ABORIGINAL FLAG (3)

Harold Thomas designed the flag because he wanted people to be liberated, to stand up for their own land rights, to be in charge of their own community and lives, and to have responsibility for their actions.

He made it bigger because he wanted all of Australia to see and recognise the Aboriginal flag and to understand what the colours stand for. Black represents Aboriginal people, red is the colour of the land, and gold is the warmth of the sun.

When I see the flag it makes me feel very proud of being an Aboriginal person. My community and family also feel strongly about the flag. The community can run its own business without European people deciding how it should be run. At the community, Aboriginal people are free to say what they want to say and to stand up for themselves. The community will change things according to what the people want. To my knowledge, Harold Thomas feels proud of what he has achieved by designing the Aboriginal flag.

The flag means to me that we are free — to be in charge of ourselves, to accept responsibility for our own actions, and to stand up for what is ours. Aboriginal people right across Australia feel and know what the flag means. It stands for the transmission of traditional knowledge, the reproduction of culture, liberation, empowerment, and social change.

Terry Sams, Peppimenarti *

*** These three articles were originally published in *Kularlaga: Journal of Aboriginal Adult Education*, Vol. 2, No. 1, 1991.**

Trespassers Keep Out, 1982 ➤
Designed by Avril Quail
Exhibit no. 23

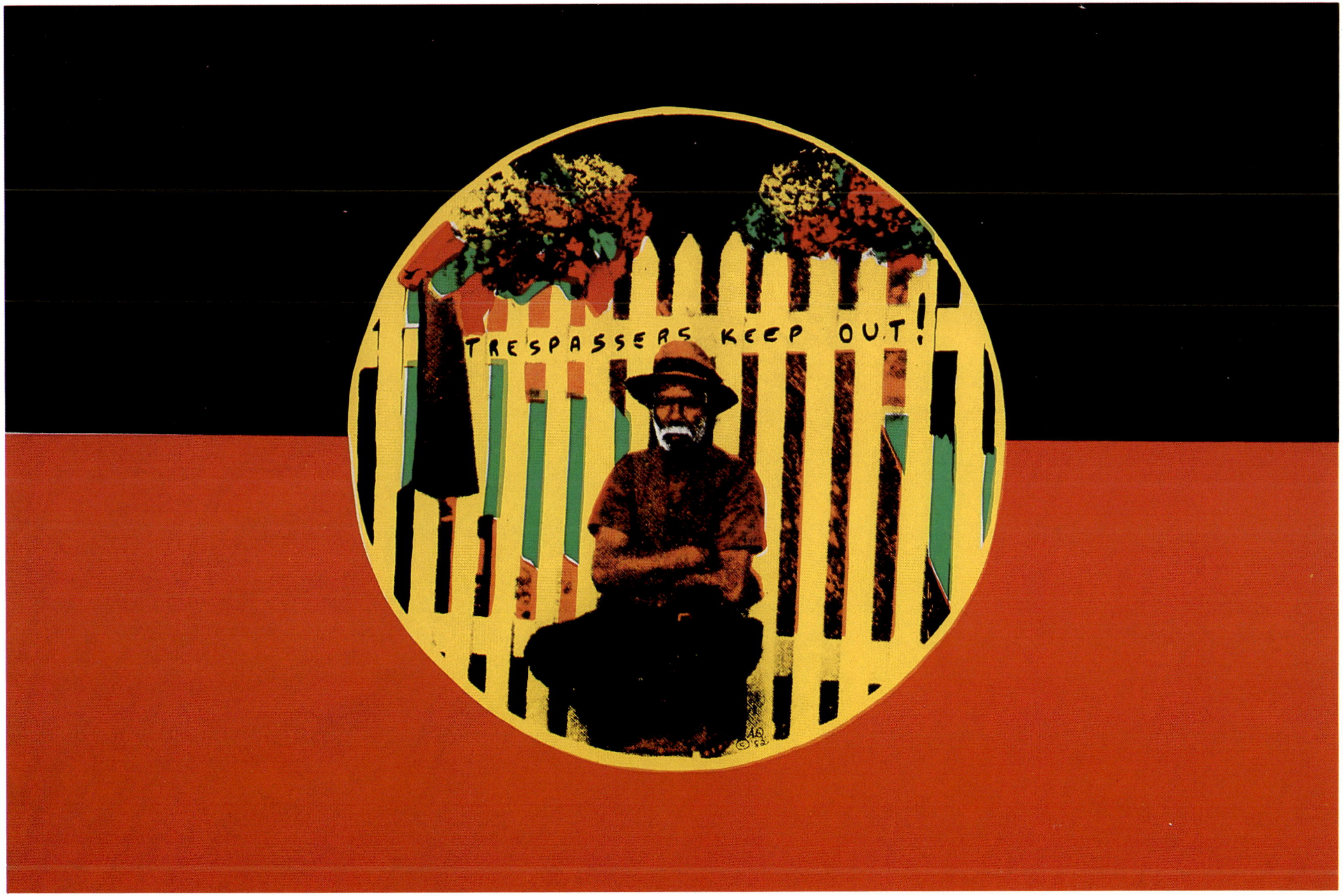
TRESPASSERS KEEP OUT!

INTERVIEW WITH MAMBO & REG MOMBASSA — Anne-Marie van de Ven

Mambo Graphics' *100% Mambo* label (now known simply as Mambo) was established in 1985 by Dare Jennings in Sydney, New South Wales. Since its inception, this Sydney-based company has drawn inspiration and images from the work of a number of enthusiastic and talented young Australian artists. Artists such as Reg Mombassa of the rock group *Mental as Anything* and Nick Seymour of *Crowded House* have created designs for screen-printed fabrics used in the production of Mambo surf fashion wear and furnishing fabrics. Many of the designs created over the last six years have parodied aspects of the Australian landscape and ways of life associated with specific environments like the beach or the suburbs.

Mambo Graphics reproduces these humorous and often bizarre images not only for its screen-printed fabrics, but also as promotional graphics which appear in local surf and fashion magazines and as designs for Mambo posters and T-shirts.

Reg Mombassa, one of Mambo's leading freelance artist/designers, was born Christopher O'Doherty in Auckland, New Zealand in 1951. Following his move to Australia in 1969, he attended The National Art School at East Sydney Technical College and then worked as a labourer, driver, musician, songwriter, and cleaner. He adopted Reg Mombassa as a stage name after he formed the band and began playing guitar, writing songs, and occasionally singing with *Mental as Anything*.

Reg's illustrations have appeared in *Vogue*, *Dolly*, *Stiletto*, *Rolling Stone,* and other magazines. His versatility as an artist, writer, designer, and musician have led him to become involved in numerous creative activities. Not only is he one of Mambo Graphics' most outstanding artists and a founding member of *Mental as Anything*, but he has also formed a new band, *Dog Trumpet*, with his brother Peter O'Doherty, and participated in several exhibitions of paintings.

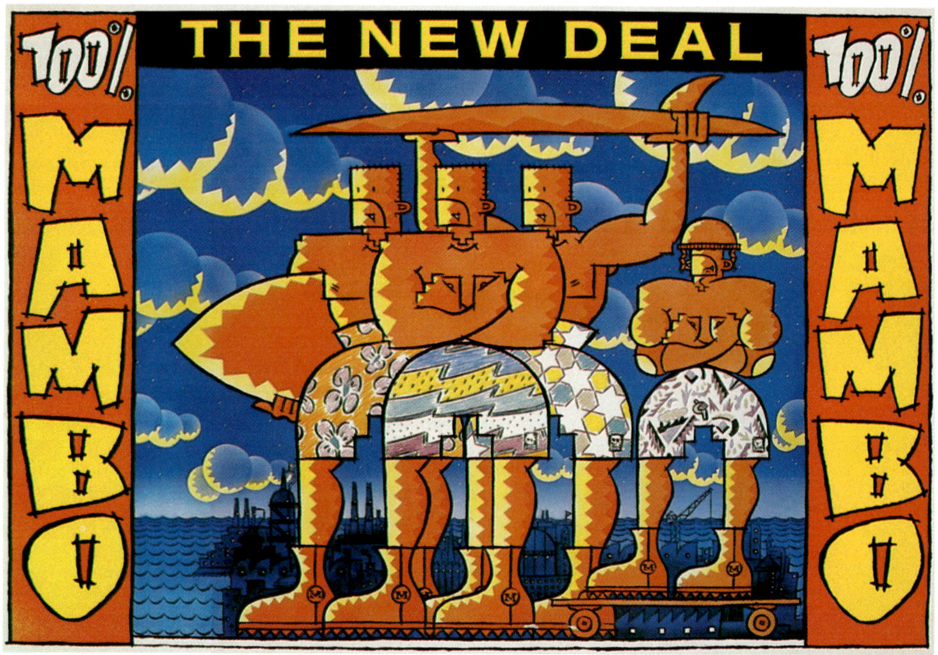

The New Deal, 1985 Parodies aspects of the Australian landscape & lifestyle Exhibit no. 22

The following is a composite response from Reg Mombassa and Wayne Golding, Mambo's art coordinator.

At what point did you begin to use Australian imagery?

Reg Mombassa: When I arrived here from New Zealand I went straight to art school and started painting local landscapes. Prior to that I used New Zealand imagery.

Wayne Golding: At Mambo we have never found it necessary to place any emphasis on directing the graphics provided by the artists. We do find on occasion a specific graphic is required for advertisements or for a poster, but, the way an image is presented is, on the whole, the decision of the artist.

How has your use of imagery changed over time and what has influenced this change?

Reg Mombassa: As I see more of the country I've included more rural and desert elements into the landscapes, also more spacemen, monsters, machines, factories et cetera. The reason for this is that as I grow older my head is being filled with nonsensical half-truths and arcane speculations, and the spillage from this metaphysical stew unfortunately affects the contents of my pictures.

Wayne Golding: As far as the graphic content goes I would say that our imagery has changed little, due mainly to the fact that the artists who were working with us in the beginning are, for the most part, the artists that are working with us today. The addition of Bruce Goold's work has certainly given impetus to that 'Australian' perception, but this is probably a question that would be better answered by the artists themselves.

Has the landscape itself been an influence on your work and what are the particular sights or elements which have been important?

Reg Mombassa: Yes. Important elements include such things as dams (circular or neatly edged), windbreaks (preferably serrated), hedges, parked vehicles, and homes sheathed in flat materials (fibro, weatherboard, cement render et cetera — the textures, colours, and associations of brick dwellings are highly unsatisfactory and can provoke

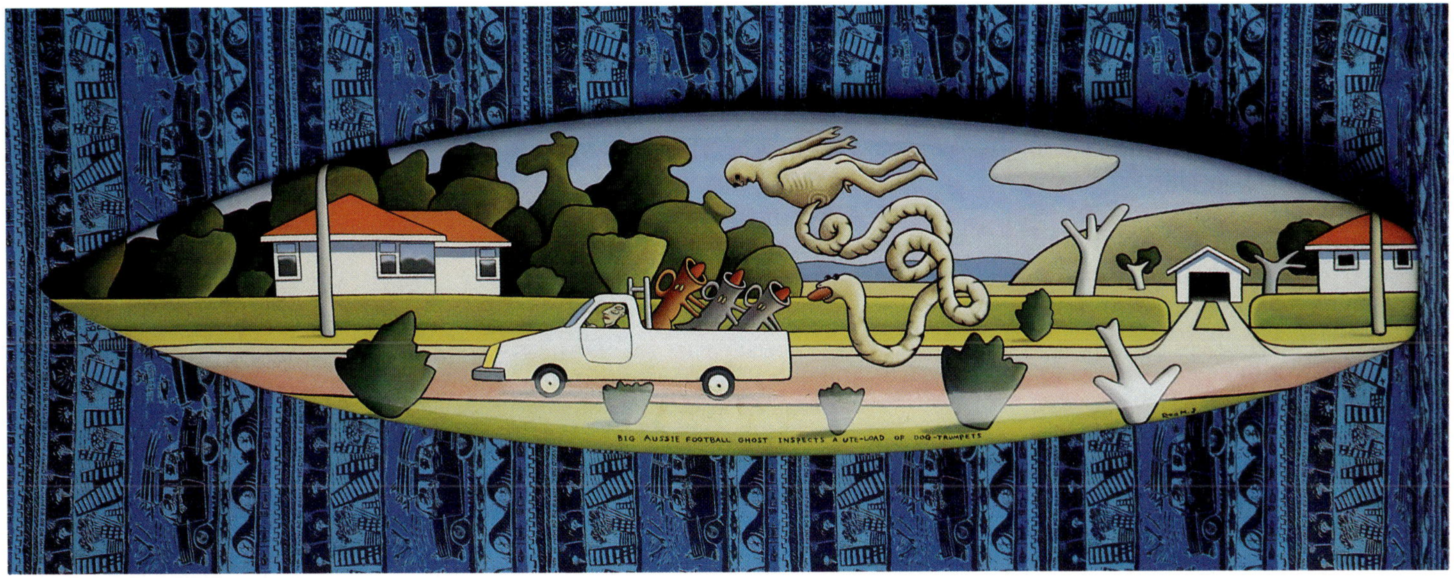

***Big Aussie Football Ghost Inspects a Ute-load of Dog-trumpets*, 1990** **Designed by Reg Mombassa** **Exhibit no. 1**

rage or minor illness if contemplated too thoroughly). Lighting and atmospherics are also very important, as I approach the landscape with the attitude of the lighting director of a small and sparsely funded community theatre group rather than that of a painter.

Wayne Golding: Certainly the influence of landscape can be seen in the works of a majority of the artists, but its mostly used as a political concern with regard to the environment. Most evident, I guess, is Reg's apocalyptical view of burning factories, trees, and houses in a surreal landscape; likewise, Nick Seymour's *Tumbling Downtown* and Gerry Wedd's car-load of lads who would never be encouraged to date your sister, rampaging through the suburbs.

I think it is inevitable that a group of intelligent and reasonably concerned artists are going to regard a landscape or cityscape in this sometimes less-than-optimistic way; being aware, as most people are today, of the potential destruction of these environments by either humankind or natural catastrophe (usually made worse by some prior human misadventure) — stripping rainforests, excessive sand mining, and soil abuse et cetera.

To what extent have you been influenced by works of earlier artists or designers who drew inspiration from the landscape?

Reg Mombassa: Probably considerably. Some of the painters whose work I like are Bruegel, Constable, Monet, Van Gogh, Rousseau, de Chirico, Munch, Lowry, Hopper, Finster. And of the locals, John Glover, Arthur Streeton, Elioth Gruner, Margaret Preston, Lloyd Rees, Fred Williams, Arthur Boyd, Colin McCahon, Peter Booth, Michael Johnson, Michael Tjakamarra Nelson, Paul Worstead, Frank Littler, and Geoffrey Searle.

Which elements of your work do you feel give it an Australian identity?

Reg Mombassa: Back-lit eucalypts, football, beer, vomiting, farm vehicles, reckless driving, gratuitous violence, and a careless sense of dress.

Wayne Golding: I think it's obvious that imagery which includes magpies on sprinklers, the ubiquitous brick and fibro house with its red-tiled roof, violent hens, and Romulus giving succour to his north coast mates would probably indicate that we are a company with its roots deeply buried in Australia.

I would also like to think that our irreverent humour makes us uniquely Australian. The Americans, of course, are likewise possessed of humour, but they are very conservative in their use of it — especially in regard to surf and street wear. Mambo, on the other hand, will physically abuse any sacred cow which is found grazing on our precinct. Our level of contempt knows no bounds and can be provided in a variety of colours and sizes.

Have these elements played a part in gaining an international market for your work?

Reg Mombassa: No. The bright colours, ridiculous subject matter, toilet humour, and slightly left-of-centre but fashionable kudos has probably contributed more to the popularity of these garments than their Australian content.

Wayne Golding: I guess there is an element of truth to that, although I also believe that it's been a long time since simply being Australian meant overseas success — *Crocodile Dundee 2* saw to that. I would prefer to think that our uniqueness of vision and the quality of the product has been responsible for the success we have had overseas . . . that and some bloody funny graphics.

PAPA, WHEN I GROW UP I'MA GOINGA TO BE AN AUSTRALIANO! — Mimmo Cozzolino

Even before I was born I knew I was going to be an Ozzie. If your name has an OZ in it, your future has been written. Anyway, I was born in Italy and came to Melbourne in 1961 at the impressionable age of 12.

I remember the sense of adventure which surrounded our trip. Before we left, my father started talking about how different Australia was from Italy. He had read about Australia and was preparing us kids for the trip ahead. He told us about the native Australians and their magical boomerangs, and the trees with flat leaves which grew pointing down. The seasons were all back to front and there was no snow at Christmas!

Above all I remember him talking about the deserts of Australia. The only desert I knew then was the Sahara and this is what I imagined Australia to be like: kangaroos hopping about sand dunes, and over the horizon Aboriginal people instead of legionnaires.

The reality was very different on our arrival at Port Melbourne. We were loaded straight onto a 'red rattler' and chuffed off to the Bonegilla migrant camp near Albury where, two weeks earlier, more than 1000 unemployed migrants had rioted. The landscape did not seem like anything I had imagined. Definitely no sand dunes, but we did see a few kangaroos and plenty of sheep.

It wasn't until much later that I experienced some of that romantic Australian landscape which had seemed so exciting to a twelve-year-old kid in Naples. I travelled to Perth across the Nullarbor Plain, and when I saw the desert for the first time, I wasn't disappointed. Then, in Ceduna, after 11 years in Australia, I met and talked to an Aboriginal person for the first time.

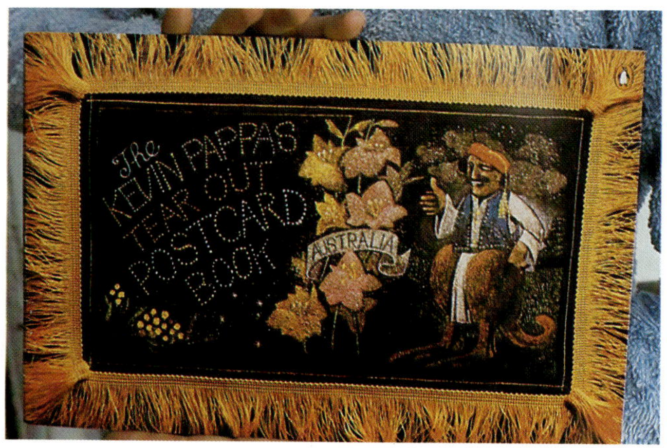

***The Kevin Pappas Tear Out Postcard Book*, 1977**
Designed by All Australian Graphics

INFLUENCES

When I was studying graphic design in the late 1960s there was no discussion about the relevance of Australian imagery as a design source. Nor was there anything about the history of Australian design. I would have loved to have known about the commercial art of people like the Lindsays, Ure Smith, Harry Weston, Douglas Annand, or James Northfield.

Instead, I knew more about what was going on in New York and Los Angeles (by subscribing to the American magazine *Communication Arts*) than I knew about what was happening in Australia. At college there was no tradition of any other kind to be passed on. We were asked to interview established local designers, but in 1970 they were mostly into Swiss or American West Coast design. It was not the sort of work that tended to inspire students.

A much more interesting influence at college was a fellow student, Con Aslanis, a Greek and a migrant like me. We talked a lot about the future and the possibility of working together in our own studio. The fact that we were both migrants really bound us together.

Con had (and still has) a great eye for the quintessential Australian suburban landscape and its inhabitants: the fluffy slippers; the 'Crappy Nappy' van; dad washing the Holden on the nature strip. This rubbed off on me. Con was also more politically aware, and we talked a lot about the changes on the federal political scene. It seemed that, after 23 years, the ALP, led by Gough Whitlam, was going to govern and bring some fresh ideas to politics: a new deal for migrants; buying back 'the farm'; pulling out of Vietnam; and making Australia more independent.

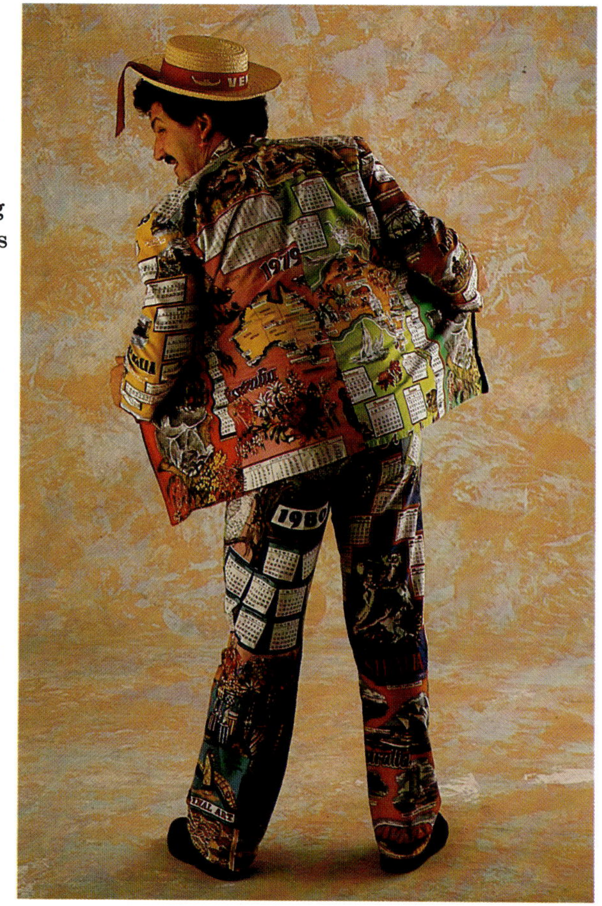

The Mimmo Cozzolino Tea-towel Suit
Tea-towel as national icon Exhibit no. 7

These were all factors which influenced us to explore an Australianness in our work, which we realised was missing in the design of the early 1970s. (I say Australianness and not Australiana because Australiana had a different meaning at that time and the word was not used the way we use it now.)

It was a time to be proud of being an Australian, especially if you were a *New* Australian. As we were about to start a design studio, we would be able to capitalise on a fresh marketing edge which no one else had: Australian graphics. After all these years I still do not know quite what this term means. Perhaps it was enough to acknowledge that if you worked in Australia, it was alright to call what you were doing 'Australian graphics'. At the time I didn't know that artists like the Lindsays, Walter Jardine, Hal Missingham, and others had done it all before.

ALL AUSTRALIAN GRAPHICS

Within this electric cultural milieu, I set up a studio with Con Aslanis in 1972 and we called it All Australian Graphics. The name was our joke — two wogs running a studio called All Australian Graphics. Our trademark was a character Con created called 'Kevin Pappas', which consisted of a Greek in traditional costume on the top and a kangaroo below.

Con and I had talked about doing postcards to promote the studio and, in 1975, we produced our first cards. They were a hit and some people even wanted to buy them. Lamella Distribution took them on and wanted us to produce more, but we also needed to sell a lot more to break even. We looked at other distributors but no one was interested in our brand of Australianness. Eventually we thought to put the cards in a book so we could utilise standard book distribution, and Penguin excitedly printed 32,000 copies. After *The Kevin Pappas Tear Out Postcard Book*, Australian cards of the modern era would never be the same!

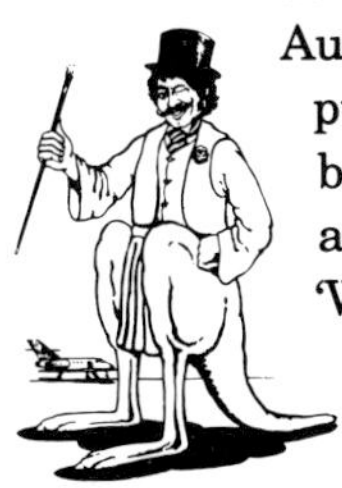

***Kevin Pappas*, patron saint of AAG**

All Australian Graffiti, as our studio was later called, disbanded in mid-1978 before we had a chance to further develop the Australianness theme. We nipped ourselves in the bud before we could offer our ideas to a broader audience. By the late 1970s Australianness was ready to be commercially exploited, but we quietly bowed out and let others carry on. Our motivation had been primarily a journey of self-discovery and of wanting to absorb our new cultural landscape. To exploit Australianness required a different kind of motivation which we just did not possess.

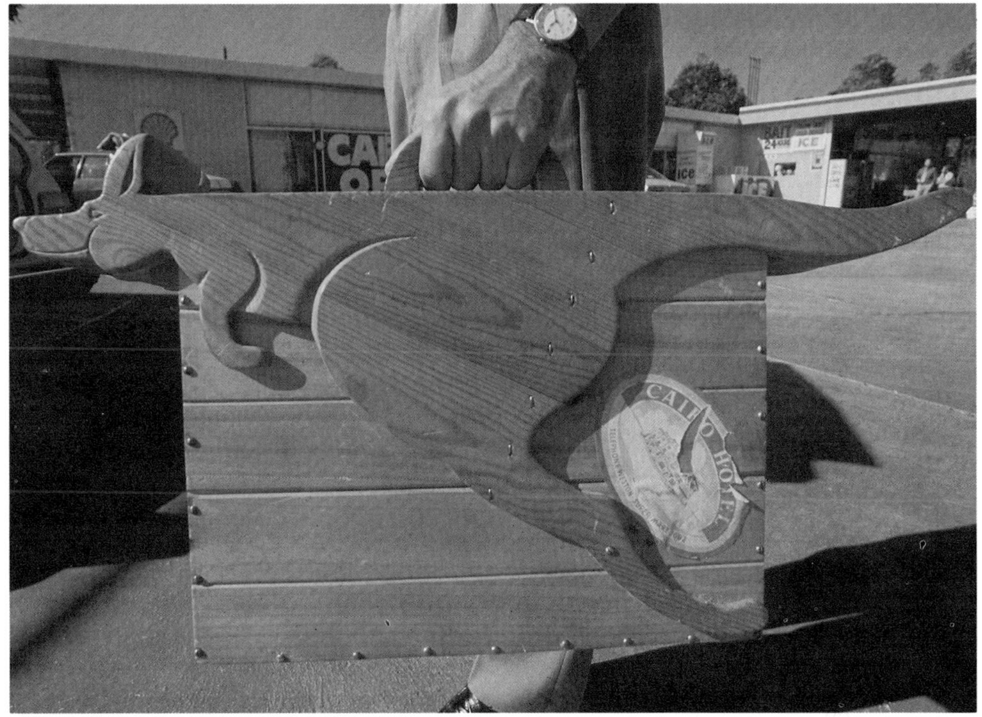

***All Australian Graffiti Suitcase*, 1975 Designed by Con Aslanis Made by Robert Reid**
To travel the Australian landscape you need the right luggage

MEDITERRANEAN CONVERTS

In retrospect, Con and I were also into Australianness in an unconscious desire to be more Australian than the Aussies. I realised this when, in 1980, I published *Symbols of Australia*. The book received a lot of media attention and the question most often asked was: 'Why has the task of publishing a book on our national symbols been taken on by an Italian? What would a *wog* know about Vegemite or Minties anyway?' My answer was that, as an outsider, I could see interesting things which had become invisible to the locals. *Symbols of Australia* is still selling and has now become a reference book. I believe it was one of the ripples that helped to create the tidal wave of nationalism which characterised the 1980s. But I have had a love/hate relationship with this book. I embarked on its research and production not really caring if I would ever get a return on the months of work I had put into it.

The love comes from the thrill of discovering, in trademarks, a culturally rich aspect of Australian commercial art never previously explored. The hate comes in being stereotyped as a designer who does only 'Australiana' design. Yet, over the years, my approach to the use of Australian imagery has not changed — I use it when and where it is appropriate and relevant to the communication.

During my 31 years here I have noticed that Australia has become more Australian than it was when I first came. I have also discovered that the harder I try to become an Australian the less I succeed.

Aroovederci!

ADVERTISING & AUSTRALIAN NATIONALISM — Gawen Rudder

I grew up in an Australia that was as remote from the rest of the world as Tibet. The tyranny of distance was combined with repressive censorship and one of the most boring social structures ever invented. Bread was white, telephones were black, spaghetti was canned, cheese was Kraft, and politics was Robert Menzies. There were very few non-Anglos in the country — a few Chinese market gardeners left over from the Gold Rush who owned restaurants or were greengrocers. By and large, we were middle class and middle-brow. We turned to Britain for the royals and our legal structure, and to America for our popular culture. Our world was a simple, if somewhat isolated, place.

In 1956 everything was to change. We had the Olympic Games in Melbourne (and the revolution in Hungary), and the medium introduced to beam our message to the world remained. The television now squatted comfortably in a corner of our living rooms and refused to be ignored. If we had hoped, however, that the mock Los Angelean accent of our radio disc jockeys was to be replaced by something more culturally relevant on television, we were to be sorely disappointed.

On Sunday, 4 November 1956, HSV–7's test pattern held all of Melbourne agog for an hour before Eric Pearce announced, in his best pommy accent, 'good evening', cut to a message by Menzies, and then proceeded with the programs of the evening. First off was *Robin Hood* from Britain, followed by Danny Webb, in his best American accent, hosting *I've Got a Secret*, followed by *Our Miss Brooks*, *Mickey Mouse Club*, *Father Knows Best* and, finally, *God Save the Queen* at 10.46 p.m.

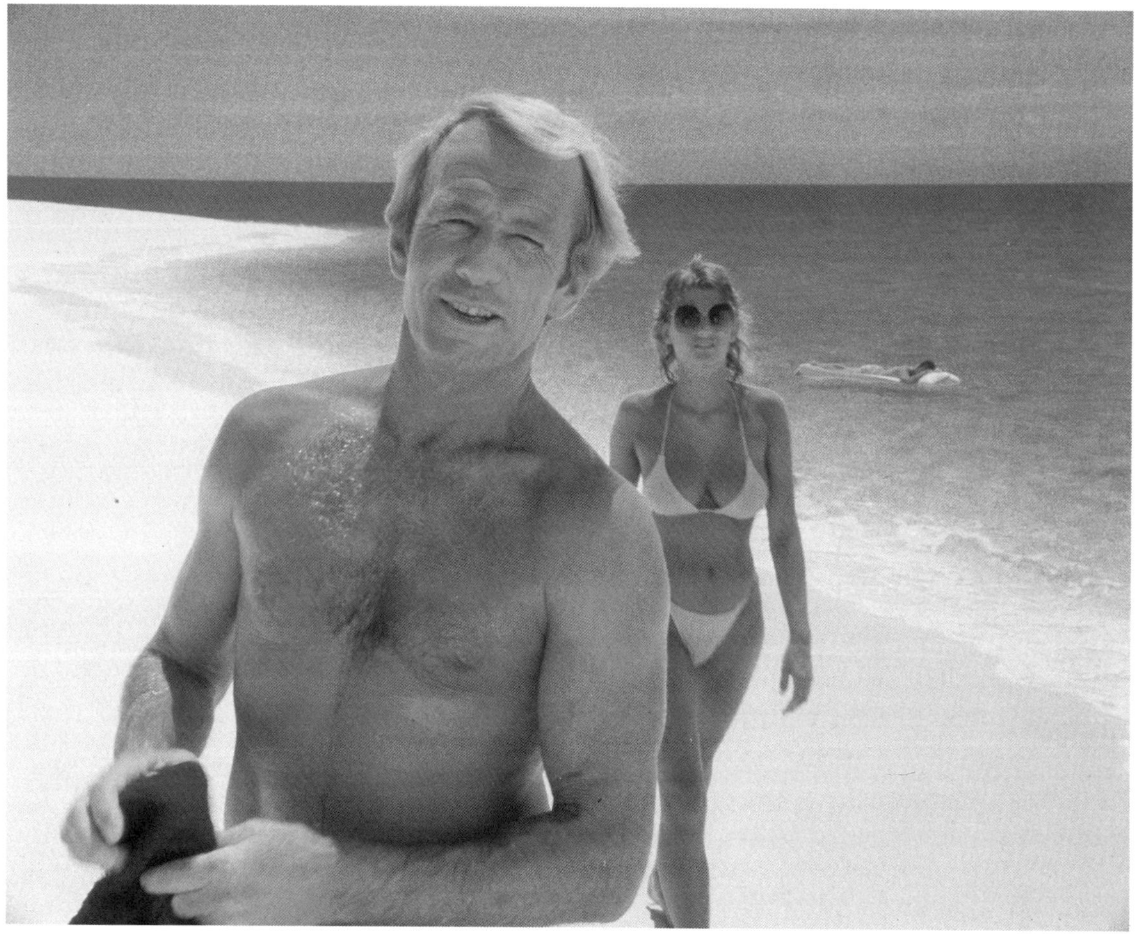

***Paul Hogan & Friend*, 1983 Photograph: Courtesy of Chiat/Day/Mojo**

The top-rating program in 1957 was *I Love Lucy*, and eight of our top ten programs were from America. In 1978, 21 years later, only three of the top ten programs were produced in Australia, prompting Phillip Adams to quip: 'There hasn't been 21 years of television, just the same year 21 times over'.

It was not until the late 1970s that things changed. By 1980 seven of the ten top-rating shows were produced in Australia, multicultural television started in Sydney and Melbourne on the 0-28 network, and the Australian accent was heard on television loud and clear from the perfectly formed lips of Paul Hogan. He said 'g'day' instead of 'good evening', and took the mickey out of both the stuffy English-style 'Benson & Hedges' ads and the American cowboy 'Marlboro' cigarette adverts with his down-to-earth Aussie 'Winfield' commercials. It took 23 years to really find our culture and accent.

In early 1983, Paul Hogan, John Cornell, and our advertising firm, Mojo, met the new Minister for Tourism, John Brown. We proposed that: 'You get the funding from the Australian Tourist Commission and we'll give you an advertising campaign that will turn Australian tourism around'.

In 1983 less than 900,000 tourists visited Australia: less than half of one per cent of the world's total. The potential for growth from such a tiny base was enormous. The Government's Economic Summit, later in 1983, identified tourism as a major potential sunrise industry for Australia. The result of the campaign, launched in January 1984, rewrote Australian marketing history.

FROM POSSUMS TO PEOPLE

Put in the simplest terms, the strategy was to switch the average American's perception away from Australia as a large, dusty land inhabited by funny looking marsupials, to a country full of warm, welcoming, friendly people. It was all about giving Australia a human face. Friendly *people*, not geography. *Australians*, not kangaroos and koalas.

Market research showed that Americans — like Australians — have a fear of being 'trapped' in a foreign culture; of being made to feel unwelcome; of being told to 'Go Home Yank'; and a fear for their personal safety (especially with the terrorist attacks in Europe during the mid-1980s). The campaign was predicated on creating a friendly, emotional, *human* image for Australia, of getting Australia in the front of impending travellers' minds, of putting our country seriously on the international tourism shopping list. It became the most successful tourism campaign ever run in America and returned $32 in tourism earnings for every $1 that had been invested in advertising.

Fifteen years ago we knew what to do with migrants, the so-called 'New Australians'. We assimilated them, which was just a polite word for homogenising them and rendering them invisible. We made them indistinguishable from Aussies as quickly as possible. We did a pretty good job too. But now, of course, assimilation is a dirty word. Now we encourage *ethnicity*. We are now officially multicultural.

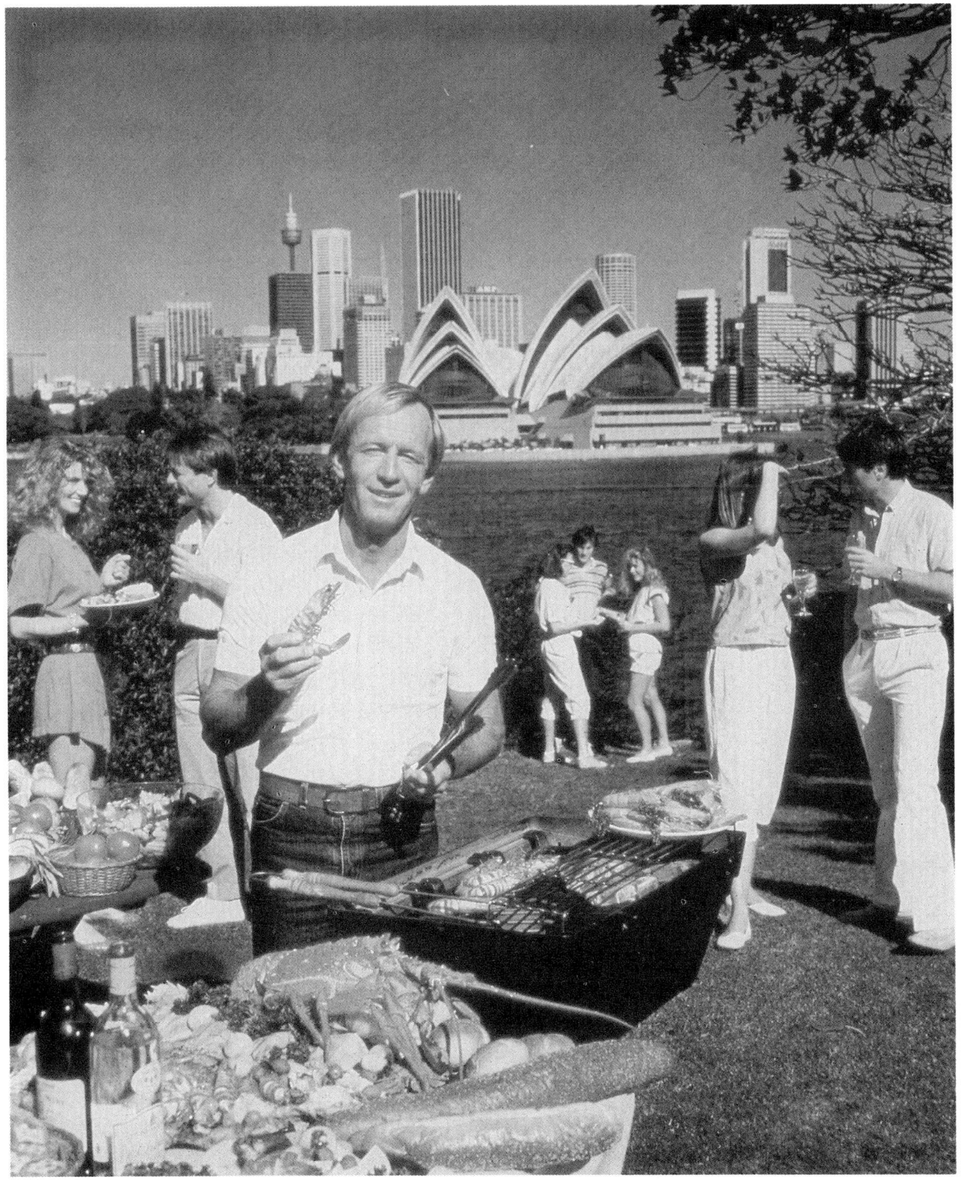

***Throw Another Shrimp on the Barbie*, 1983**
Photograph: Courtesy of Chiat/Day/Mojo

And how do you think people feel about that? Some feel very nervous indeed. They feel as though their sense of Australian identity is under threat. And, of course, it is! They are being called upon to re-define what it means to be an Australian.

It is hard to imagine doing anything to a culture that would make people feel more insecure than by telling them their national identity is up for grabs. Is it really surprising that racism is on the rise again? Is it any wonder Australians are taking refuge in myths, stereotypes, and legends from the past — romanticising images of the bronzed ANZAC and the Man from Snowy River? We still equate these simplistic images with what it *used* to mean to be an Australian. We are still thinking in the past, protecting ourselves against present and future realities by indulging in rose-coloured sentiment.

Today, 35 years after the launch of television, our relationship with the medium has changed. Our television is now more intrinsically Australian than ever before: Australian soapies and their stars, sports coverage, 'elder statesmen' from Brian Henderson to Molly Meldrum, and advertisements such as Louie the Fly and Vegemite have acquired a comfortable familiarity.

We have adjusted our sights and lowered our expectations. The bright fires of nationalism have become warm embers. All the easier now to drop off to sleep in front of that television set which once refused to be ignored.

INTERVIEW WITH KEN DONE — Anne-Marie van de Ven

Ken Done was born in Sydney in 1940, and spent a large part of his youth in Maclean on the north coast of New South Wales. Later, he studied at the National Art School in Sydney and worked for 20 years as an advertising graphic designer in Sydney, New York and London, before turning his energies to painting. His works present Australian landscapes, flora, fauna, gardens, and beaches, with many inspired by the corals, fish, and waters of Australia's east coast.

After holding his first one-man exhibition in 1980, Done began translating his most successful images onto clothing, beachwear, calendars, and bed linen. (Since 1988 this aspect of design and production has been undertaken by Judy Done.) The products are made by various local and overseas manufacturers and sold through Ken Done outlets and major department stores.

With the launch and success of the Ken Done product range, Done's drawings and pictures of the Australian landscape have become familiar images in the minds of a vast proportion of the population, not only in Australia but also overseas. One popular weekly magazine in Japan, *Hanako,* which targets women aged between 18 and 35, has featured Ken Done designs on the cover of every issue since 1989. These designs include some of his wonderfully bright and colourful Barrier Reef series, a series which evolved from years of sketching and painting in ink, crayons, coloured pencils, and oil and acrylic paint, on paper and canvas.

Were you encouraged to use Australian imagery during your art training ?

I attended the National Art School between 1954 and 1959. As my boyhood had been spent on the north coast of NSW, I had already taken in, and was familiar with, the imagery of the beach, the river, and the coastline. This, in fact, constituted my playground.

At art school, the earlier years were spent studying the classic skills, and it was really through direct contact with Australian artists such as John Coburn, John Passmore, Frank Norton, Strom Gould, and Phyllis Shiletto that we were given access to Australian images.

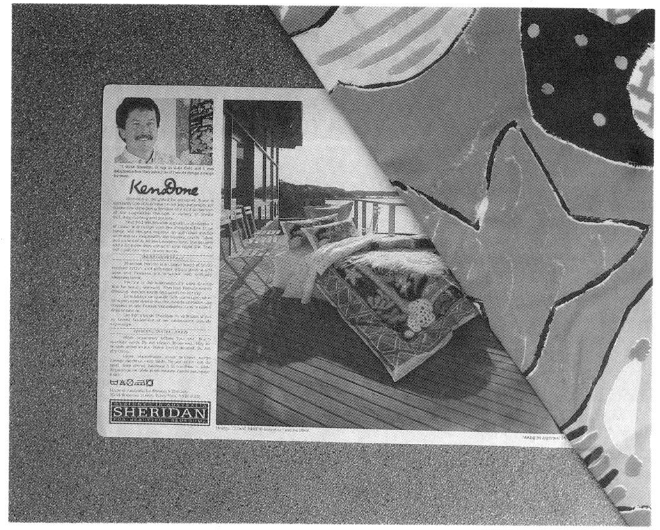

Doona cover, 1985 by Ken Done Inspired by Australia's coast

How has your use of Australian imagery changed over time and what has influenced this change?

The love affair I have had with the beach, the Barrier Reef, the flowers, the excitement of the Australian outback, will, I am sure, continue all my life. I am sure some of the imagery made of the beach, for instance, could not have been done unless I had been influenced by Fred Williams. Equally so, every time one attempts a drawing of the bush, it is possible to feel the magic of a Hans Heysen, Albert Namatjira and Sidney Nolan.

The changing use of the imagery simply comes from the artist's continual quest for new or refined ways of communication, although I personally feel it's important not to lose sight of the fresh simplicity you feel as a child.

Has the landscape itself been an influence on your work and what are the particular sites or elements which have been important?

The immediate landscape of my studio has been a very dominant feature, as have been the many parts of Australia and the world I've been fortunate enough to visit. Every time I've been diving on the Barrier Reef or reefs in the South Pacific, I am inspired afresh to try to produce works which convey the beauty of that underwater environment. Each day at my studio, overlooking Sydney Harbour, I find a constant source of inspiration, whether it be the cool, soft mauve of winter or the bright cacophony of the beach in summer.

How have you been influenced by works of earlier artists or designers who themselves drew inspiration from the landscape?

I have been influenced by works as diverse as sand painting, where the landscape itself becomes the palette, or early pioneers' more naive views of Australia, or the rich imagery of Arthur Boyd, Sidney Nolan, Albert Tucker, Albert Namatjira, Lloyd Rees, and the photography of Max Dupain, David Moore, or Rennie Ellis. All artists, I believe, are influenced by other artists. It's finally the way in which you express your own feelings that is important, but it's necessary to acknowledge your debt to the past experiences of others.

Your work has been described as Australian. What elements of your work give it this identity?

In my personal work there were some earlier 'images' — Sydney, the Harbour, the beach — that were unmistakably Australian. My first Sydney Harbour T-shirts, for instance, were really the same drawings I'd produced on paper for my exhibition, reproduced on the medium of a T-shirt to remind people of the exhibition. The fact that this, and other images about Australia, seemed to strike such a chord was only possible, of course, because a large number of Australians were proud of their Australianness.

TWO SIDES TO TOURISM — Galarruwy Yunupingu

Tourism in the Northern Territory is big business. In 1986 it was worth $250 million to the Territory economy and the figure has been steadily growing.

What brings tourists to the Northern Territory? The main attractions are not the casinos. Visitors come to experience our landscape, the outback life generally, and especially Aboriginal culture. Surveys show that at least a third of visitors to the Northern Territory come here expecting to experience something of Aboriginal culture.

Although Aboriginal culture, our land, and the wildlife (which are our totems) are very important tourist attractions, Aboriginal people hold a very small share of the tourist industry. I believe, however, that this share will increase and, through involvement in tourism, Aboriginal people will be helped to become more economically independent. Even though our share so far has been small (after all, the Government has not been offering millions to us to build big resorts and casinos), there are a number of very important tourism areas where our people are already working.

ACCENTUATE THE POSITIVE

Currently, there are almost 70 Aboriginal people employed as rangers for the Australian National Parks and Wildlife Service and the Northern Territory Conservation Commission. There are a number of good people in the field who want to run the parks with strong Aboriginal involvement, both in the areas of long-term management and day-to-day park services. This is not helped, however, by politicians using the parks issue as a political football. Not only do Aboriginal people get tired of being kicked up and down the field, but this behaviour discourages tourists from visiting the Territory.

Seeing Aboriginal people as owners of park areas and working in the management of parks is something special and makes the Territory an interesting place to visit. This is true especially of Kakadu and Uluru. If the Jawoyn people's land claim is successful and they become owners of the Katherine Gorge National Park, this will help the growth of tourism at Katherine in a way that is beneficial to Aboriginal and non-Aboriginal people.

The operation of Government park services is only one of the areas in which Aboriginal people are working. In Central Australia the Aboriginal community at King's Canyon is very much involved in the private development which is taking place there. They are becoming involved in joint ventures with such experienced tour operators as Bill King's safaris.

Through paintings, carvings, and special designs on weapons and tools, Aboriginal people record their stories — just like writing in a book. However, the financial side of our art and crafts is important too. There are about 2000 Aboriginal people in the Northern Territory who produce art and craft for sale. They receive about $2 million a year for their work, but the value of their work to the Territory economy is, in fact, much higher. Using the methods of the Northern Territory Government to estimate the value of tourism, Aboriginal arts and crafts are worth more than $4 million a year to our economy.

Cooinda Hotel-Motel is another example of the way Aboriginal people are becoming involved in tourism. There, the Gagudju people are developing the Northern Territory economy by channelling the money earned from mining on Aboriginal land into tourism.

ELIMINATE THE NEGATIVE

Of course, there are bad, as well as good, sides to tourism. For one thing, Aboriginal people, like most people, do not like being stared at by strangers or asked silly questions.

We are also concerned about sacred sites. Many are considered dangerous. If they are damaged or if strangers go into the places without knowledge or respect, the owners and custodians can become sick. It is also possible that the people who have gone there, or others, can also be harmed. When I go to a strange place I know that I must be careful and follow the rules for that place which the owners and custodians know. This is the law that we have always followed.

Concern for these special places is one of the main reasons why we want to ensure that tourists are looked after properly when they come onto our land. Our respect for these places is something that we want other people to share. It is the reason why Aboriginal control of tourism on our land is so important.

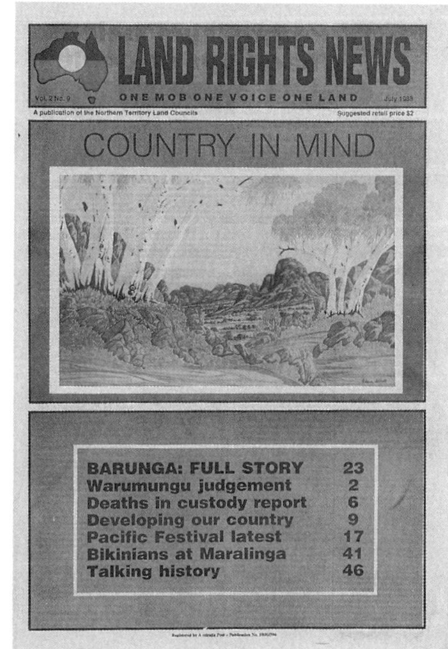
LAND RIGHTS NEWS

ONE MOB ONE VOICE ONE LAND

COUNTRY IN MIND

BARUNGA: FULL STORY	23
Warumungu judgement	2
Deaths in custody report	6
Developing our country	9
Pacific Festival latest	17
Bikinians at Maralinga	41
Talking history	46

***Land Rights News*, 1988 Cover painting by Clem Abbott Exhibit no. 56**

While we are going through this time of learning to respect each other's culture, it is important there are sensible controls to see that the customs of both cultures are not abused. If I am in your backyard you have a right to expect that I will respect your wishes and concerns about my behaviour . . . and when you are on our land we expect the same.

Reprinted from *Land Rights News*, Vol. 2, No. 5, December 1987.

NB Since publication of the original article, the Jawoyn people have been successful in their land claim.

The advent of mass tourism in Australia in the 1880s coincided with the development of colonial railway networks and increased standards of living for middle and working class Australians. Private transport companies and public agencies began to publicise and promote the natural attractions and locales that could be reached by means of the transport services they offered. Ever since, coach and steamboat companies, railways, automobile clubs, and airlines have promoted seaside resorts, mountain retreats, lakes and estuaries, underground caves, and rivers and gorges as desirable destinations for tourists and travellers.

The first state-operated tourism information service in the colony of Victoria was established in 1888 when the Railway Commissioners appointed a Railways Inquiry Officer to provide information to people visiting the city for the Melbourne International Centennial Exhibition. Similarly, in Sydney, the first official tourist guide, the *Railway Guide*, was produced in 1879 by the New South Wales Railways to coincide with the Sydney International Exhibition.

As the railways remained the principal means of conveying tourists until the 1950s, when the motor car and the aeroplane became the popular modes of domestic transport, it is not surprising that their publicity departments produced some of the more interesting and evocative literary and pictorial descriptions of holiday resorts and the natural landscape.

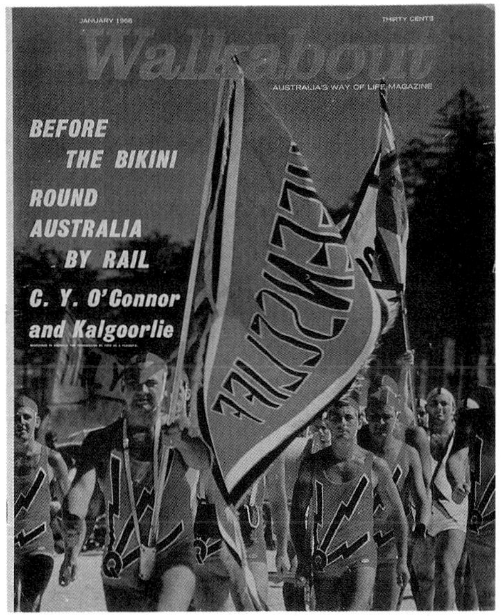

***Walkabout*, January 1968 A key image of beach life Copyright: Australian Tourism Industries Association**

The quaintly titled *Marine Health Resorts: Cheap Trips by Rail from Sunbaked Plains to Seaside Nooks*, published by the Victorian Railways Department in 1902, was produced:

with the dual object of encouraging traffic and promoting health . . . [for] the Railway Commissioner and his officers devote special attention to the carrying of country residents from the torrid interior to the invigorating air of the seaside.

The civic-minded Commissioner and his staff reminded their readers that a vacation during the summer months, after the harvest, should not induce feelings of guilt in the minds of hard-working toilers of the soil, for:

January, February and March are . . . specially suitable for a healthful pilgrimage to marine resorts, the cost of which is saved tenfold in doctor's bills and renewed strength to pursue the arduous toil which is the lot of those on the soil who are the basis of our prosperity.

Country folk had a duty to leave their properties and take a holiday courtesy of the Victorian Railway Commissioner, for the good of the community. A noble piece of promotional writing.

For water-borne journeys, Edwardian travellers and tourists took coastal or river steamers. Privately owned transport companies, such as the Sale Steamboat Company which operated a steamer service on the Gippsland Lakes, also published tourist guides for the public. One publication, *A Guide to Gippsland Lakes and District* (Sale, c. 1910, originally published c. 1880), includes an essay titled 'A Lady's Impressions of the Gippsland Lakes'. The writer describes with a rapturous, poetic enthusiasm the beauties of the lakes which lie about 200 kilometres east of Melbourne. Her enthusiasm for the lakes, however, is conditional, for in part their beauty lies in the remembrance of other, European landscapes:

The very boats suggest Scotland. Only the black swans and pelicans are there to dispel illusions. Later on the carol of magpies and the glory of the wattle brings us finally back to Australia, and we leave the Lakes behind us firm in the conviction that neither Scotland nor any corner of the Empire can claim the monopoly of beautiful waterways and unforgettable sweeps of country.

A tendency to regard our landscape as beautiful as far as it approximates the European landscape did not cease with the passing of the Edwardian period. As late as 1958 the *Pocket Year Book of Tasmania* praises the charms of the Apple Isle:

Its attractiveness lies in its natural beauty consisting of mountains, lakes, rivers, fertile valleys, its sea girt shores and the charms of its cities and towns; all of which are reminiscent of the British Isles.

AUSTRALIA GOES *WALKABOUT*

Between the two world wars tourism slowly became more institutionalised, reflecting, in part, the gradual emergence and official recognition of the car as a popular means of transport. During this time of increasing awareness of the automobile and of touring by car, *Walkabout* first appeared. Launched in 1934 from the Australian National Travel Association in the Flinders Street Railway Building in Melbourne, the new magazine's stated purpose was for:

Australians and people of other lands . . . [to] learn more of the vast Australian continent and the colourful islands below the equator in the Pacific (November 1936).

Interestingly, though, during this decade, the magazine had a distinctively foreign and exotic tone to it. The articles were almost exclusively written about overseas or British Empire holiday destinations, in such a way as to excite the travel appetites of society debs and matrons, and to entice them to Britain first, and then possibly Europe or the United States by ship or a combination of ship and seaplane.

The monthly issues of *Walkabout* in 1937 included feature articles covering topics on Edward VIII's exile in Austria, and the delights of post-imperial Vienna and Salzburg (March), seaplane voyages from San Francisco across the Pacific to Sydney (April), the palaces of London, and the architecture and tourist sights of Morocco (May), skyscrapers in New York (September), and the Vatican and Tristan da Cunha (October).

Interspersed with the foreign articles were, by comparison, more mundane and matter-of-fact ones written about Australian themes: for example, tuna fishing off the New South Wales coast (February), building the transcontinental railway (March), farming 113 million sheep (May), the Northern Territory overland telegraph line (July), and a romantic account in October by Ion Idriess of pearling in the Coral Sea and off the coasts of New Guinea and northern Queensland.

The images of Australia were not primarily of the natural landscape, but a backdrop to practical, more manly pursuits such as farming, building roads, shearing sheep, or carting great loads across deserts. This preoccupation with the landscape as a repository of resources to be utilised by men, and of development and settlement, is a marked feature of the articles that appeared in the next two decades.

WALKABOUT GOES AUSTRALIAN

Walkabout became a more inward-looking, Australian-oriented magazine with the start of World War II. It changed entirely into a magazine of Australia — of our flora, fauna, resources, geography, cities and towns, and natural landscapes which featured as never before.

Because of the war, people did not travel abroad, and by concentrating on Australia, the magazine served the nation's security and propaganda needs: the wholesome, homespun topics which were published focused the attention of readers on the values and worth of the homeland, and not on the outside world and the threat beyond Australia's shores. An apt example of this was an article published in April 1941 on the 'palaces' not, as in May 1937, of London, but of Melbourne, with a photo essay on the Treasury Buildings, the Houses of Parliament, and Bishopscourt.

Mention of the outside world was in the form of travel and photo essays on Papua New Guinea, New Zealand, and our other near neighbours, and allies, in the Pacific and Indian oceans. Among the front covers for 1941 were photographs stressing masculine and martial imagery: tall, powerfully built lifesavers; an earnest, young farmer and draught-horse; New Guinea warriors with war paint and weapons; and Aboriginal youths canoeing.

The focus of the magazine in the 1950s was still Australian, with some mention of the Pacific and New Zealand and the recurring themes of development and settlement. There was a hint also of a growing interest in natural beauty and landscapes: Thursday Island and Lake Pedder feature in a 'Camera Supplement' in April 1956.

In 1962 the editors announced a new format. *Walkabout* was then printed in offset with more colour, more photographs, and regular features by authors such as George Johnston, Keith Dunstan, Myra Roper, and Charmian Clift. As well as standard pieces on poisonous snakes, lyrebirds, Sydney Harbour, and historic Tasmania, the editors showed a growing concern for cultural topics, with two features in 1964 on the new Adelaide Festival (February and November) and various Melbourne theatre companies (July).

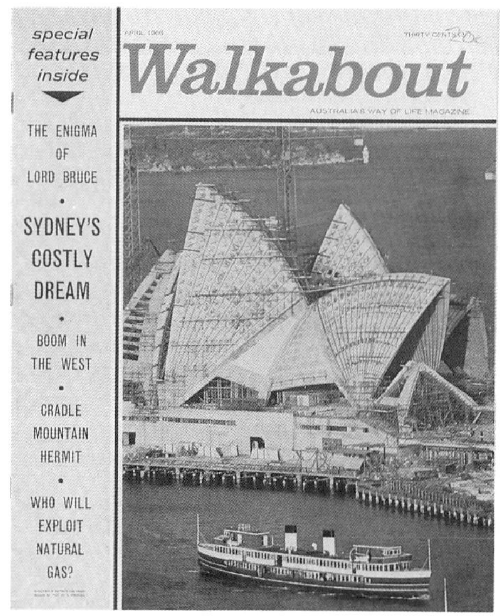

***Walkabout*, April 1966** **The Sydney Opera House soon supplanted the Harbour Bridge as our best known symbol** **Copyright: ATIA**

Today marketing consultants, tourist promoters, and travel agents advise that foreign tourists come to Australia to see only three sights (sites) — the three 'Rs' of tourism, 'the Reef, the Rock, and the 'arbour' — all in ten days. All three are natural attractions.

In support of this conventional wisdom, the Australian Tourism Commission (ATC) devises marketing strategies for European audiences which portray this country as a land of 'soft adventure' (in particular, for Scandinavians and northern Europeans). Television and print advertisements used in the ATC's European campaign project a clear image of the sunburnt country and use familiar icons such as the Olgas, Ayers Rock, Sydney Harbour, and vast sandy bays and inlets (*ATC Annual Report*, 1990).

Travellers and tourists in Australia have always been attracted to notable features of the natural landscape and have visited by car, rail, and foot, our escarpments, waterfalls and lakes, mountain retreats, beaches, and deserts. And they were doing so long before enterprising businesspeople set up amusement parks on beach and harbour frontages, or began to build the big banana, giant earthworm and big merino which have appeared in cities and towns throughout Australia.

In Victoria in 1985 a parcel of public land called the Langwarrin Military Reserve was renamed the Langwarrin Flora and Fauna Reserve, and some of the signs of its early military occupation were removed or obscured. The reserve, one of Victoria's oldest military camps, was established almost 100 years earlier in 1886 and, in the course of its use for military training and as the site of a prisoner of war camp and venereal disease hospital, the land was extensively cleared and grazed. In 1948 the Langwarrin and District Progress Association proposed 'that this little Park of 20 acres should be left as a Memorial reserve in honour of old soldiers', but the idea was never followed up. By 1975 this island of remnant and regenerating vegetation was being valued in a different way.

An ecological survey of the reserve was commissioned by the Cranbourne Shire Council and, a decade later, the land emerged under its new name as a sanctuary for nature conservation. Some large pine trees, evidence of the original settlement, were removed. The author of the ecological survey, Winty Calder, had become fascinated by the area's military history, which she saw as integral to its natural history, but her book about its past could not be launched in the newly declared reserve. Perhaps this was due to a lack of toilets or catering facilities. Or it may have been because, as one senior land manager put it, 'there's a very conscious policy not to acknowledge history'. Park administrators were 'quite terrified of the power of pictorial evidence' that showed clear paddocks pegged out with military teepees. The history carried unwelcome and contradictory messages for a flora and fauna reserve: that this bushland had been 'disturbed', that much of the site of this 'original remnant of vegetation' had been previously cleared, and that abused bushland could regenerate quickly.

This abbreviated tale of Langwarrin raises a number of issues, both positive and disturbing. The story alerts us to the fruitful collaboration that can easily arise between the disciplines of ecology and history in environmental studies. But, at the same time, our attention is drawn to an apparent conflict between the movements to preserve natural and cultural heritage.

The metamorphosis of Langwarrin is only one of many examples from Australia and abroad that illustrate the marginal, and even oppositional, status of history in the modern evaluation and interpretation of perceived natural landscapes. This conflict is not new, but it is more sharply defined today. It is a result of two developments, both of which have accelerated in Australia since the 1960s: the dominance of ecological criteria in the assessment of environmental values, and the broadening of our historical perception of landscape from isolated sites to whole cultural patterns. In much of Australia it is translated into a daily management issue by having the responsibility for 'historic places' vested in state government departments primarily devoted to nature conservation, a case of bleeding sepia into green.

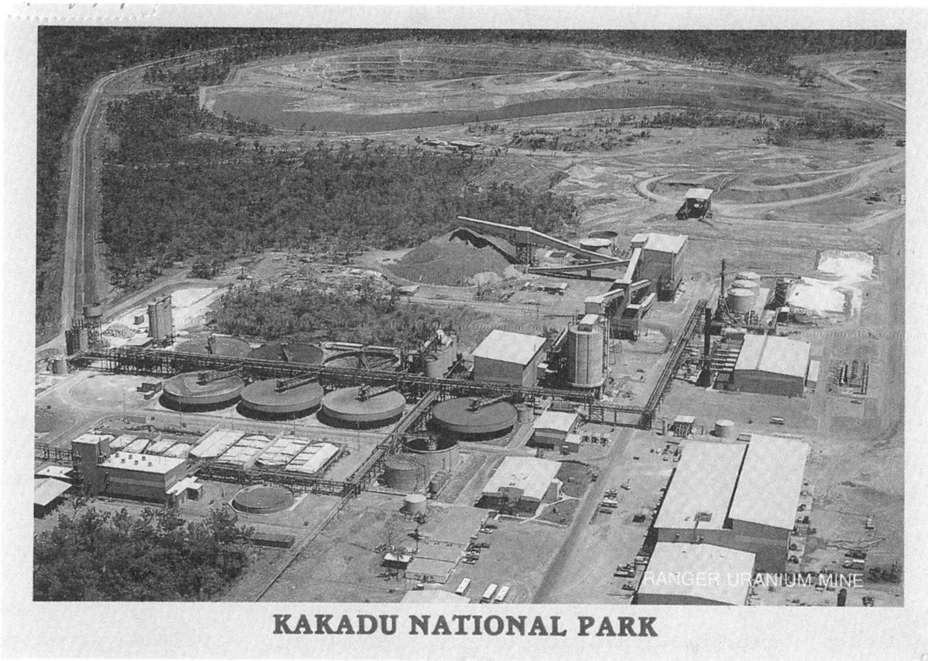

Ranger Uranium Mine **Souvenir postcard of Kakadu National Park reproduced with permission of North Australian Trading Company**

It is part of the 'green' aesthetic to separate, and sometimes eliminate, culture from nature, and 'wilderness' has become one of its simplest and most popular manifestations. Although apparently biocentric in rationale, the modern wilderness movement presents an historical vision of landscape that is unacknowledged and problematic, that is as much about humanity and aesthetics as it is about science.

Many of the persisting definitions of wilderness are sensory: they concern boundaries of sight, lengths of walks in one direction, and the maintenance of an authentic sense of human danger and isolation. Wilderness, wrote the American forester Aldo Leopold in 1921, was 'a continuous stretch of country preserved in its natural state, open to lawful hunting and fishing, big enough to absorb a two weeks' pack trip, and kept devoid of roads, artificial trails, cottages, or other works of man'. Early Australian definitions of wilderness echoed the American version: a wilderness was somewhere 'that one may be able to travel on foot in any direction for at least a full day without meeting a road or highway', wrote the New South Wales bushwalker and 'father of wilderness', Myles Dunphy.

These are primarily negative definitions, seeing wilderness in terms of remoteness and absence. They suggest how we might come to see wilderness as capable of being 'restored' by removing 'incursions' and 'distractions'. In the Virgin Islands, all traces of seventeenth- and eighteenth-century cultivation were expunged from a new national park so as to restore a 'wilderness' landscape. Modern wilderness photographers enact this vision in their choice of frame, omitting the eroded path that led them to their view. In a recent *Descriptive Report* on wilderness in Victoria by the Land Conservation Council (1990), historic relics (like maintained structures) were seen as negatives in the evaluation of 'aesthetic naturalness'. And if the definitions are about the feel and look of a place, 'wilderness' does not need to be ancient, pristine, and timeless; only to seem so.

In 1987 the eastern coast of Wilson's Promontory in Victoria, an area with a history of sealing, logging, grazing, mining, and settlement, was declared a zone of high wilderness quality. In Britain 'wilderness' has come to mean the domesticated and ecologically poor landscape of the moors. But, as Marion Shoard has explained, 'most moorlands are relatively recent landscapes, created at most 4000 years ago through the destruction of forest to provide wood, charcoal, or sheep runs'. The moors rely on burning and grazing for their continued existence, but offer loneliness and a primeval appearance to visitors.

Restoring and maintaining wilderness, even if it denies history, is a complex historical enterprise. These areas recreate landscapes which we believe existed at an earlier time. In Australia that time is, symbolically, 1788. Wilderness zones are the Sovereign Hills of the bush. They are preserved, managed, restored. They attract tourists, offer the feel of the past, and commemorate and mourn what we have lost. The trick is that these are not primeval, non-human landscapes. As Thoreau put it in 1859, 'What we call wilderness is a civilisation other than our own'. In Australia, in Sylvia Hallam's words, the land 'was not as God made it. It was as the Aborigines made it'.

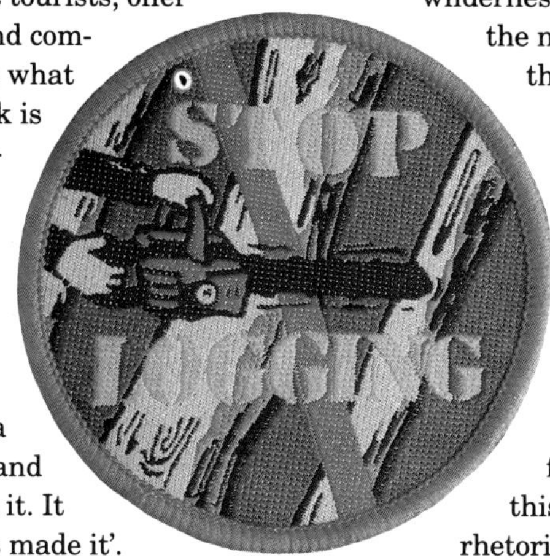

The wilderness idea is not just anthropocentric, it is Eurocentric. It preserves or restores landscapes as Europeans supposedly found them — and as Aboriginal people made them — and it calls them untouched, pristine. Aboriginal people are thereby rendered invisible as agents in the landscape. Is this *terra nullius* in another form? Just as in nineteenth-century museums, Aboriginal people are classified with nature. Only now it is done with reverence, for they are no longer Primitive Savages; they are Ecological Beings.

Although appreciation of Aboriginal manipulation of the environment is increasingly a factor in modern park management, the old language and dismissive stereotypes still slip into the new environmental handbooks. Terms such as 'untamed wilderness' appear in the first published draft of the *Kakadu National Park Management Plan* (1980). The recent *Descriptive Report* on wilderness in Victoria surveys the new scholarly insights that reveal the long and active engagement of Aboriginal people with a changing landscape, but concludes that 'before the arrival of Europeans, however, all of Australia was essentially unmodified'. What can explain this slippage in wilderness rhetoric? First, environmentalists sometimes have an uneasy relationship with archaeologists. Second, where the past occupation by Aboriginal people draws attention to a history of massive landscape change, or where their present land rights claims threaten future change, wilderness supporters have confronted political and philosophical difficulties. Both of these sources of uneasiness were evident in the battle to save the Franklin River in Tasmania's south-west in 1982–83.

Archaeologist Rhys Jones has postulated an intriguing human history of south-west Tasmania. There, 20,000 years ago, the most southerly humans in the world hunted marsupials in an open tundra. The forest as we know it today was not there: it was confined to the sheltered river valleys. These people occupied the inland valleys until perhaps 12,000 years ago when, as the glacial period waned and the rainforests expanded, they and the animals they hunted abandoned the region. In quite recent times, just before the arrival of Europeans, Aboriginal people had begun to explore or colonise the south-west again. The archaeological evidence of this human history of the forest and tundra became critical in the protection of this World Heritage Area. According to former Prime Minister Malcolm Fraser, it may have been the most important factor in the saving of the Franklin.

AN UNEASY ALLIANCE

Three main groups opposed the plans of the Hydro Electricity Commission (HEC) to flood the Franklin. Those three groups — Aboriginal people, archaeologists, and environmentalists — formed an uneasy, shifting alliance. The tensions within that alliance may be more interesting and significant in the long-term than the conflict the groups had with the HEC. The sensitive relationship between archaeologists and Aboriginal people over the excavation of sites has been widely discussed, but the other two sides of the triangle have received less attention.

Archaeologists working in the south-west found themselves and their discoveries marginalised by many supporters of the Wilderness Society. John Mulvaney has written that:

As an archaeologist involved in the Gordon–Franklin dam debate, I initially found marked lack of interest in the Aboriginal prehistory cause from many environmental conservationists. The fact that humans occupied south-west Tasmania during Ice Age times, *before* the forests developed their impenetrability, perhaps ruined their idealistic concept of a timeless *Wilderness*.

Some environmentalists called the archaeologists 'mind merchants' and accused them of 'disturbing the forest' with invasive technology and science.

There was, at times, 'open antagonism' between the Tasmanian Aboriginal Centre and the Wilderness Society. Some Aboriginal people were openly critical of what they saw to be the Society's hypocrisy: that it was prepared to use the existence of early Aboriginal occupation of the south-west to support preservation of the Franklin, yet was reluctant to actively support land rights claims. Tasmanian Aboriginal, Jim Everett, claims:

The wilderness mob continually wanted to push us to the back, to quieten our struggle and keep conservation alone up-front.

Kutikina Cave became symbolically important to both of these debates. Rediscovered in 1977, it is a huge, deep limestone cavern and one of the richest archaeological sites in Australia. Its occupation has been carbon dated to 15-20,000 years ago. For the HEC, its timely discovery seemed trumped up. Even its first name, Fraser Cave, after the then Prime Minister, smacked of cynical politics. For Rhys Jones, after 18 years of searching, it was 'the site I have been looking for . . . '. For Tasmanian Aboriginal Michael Mansell, 'it was like coming home', and 'the most important cultural thing that's ever happened to us'. For Jim Everett it was a satisfying 'spanner in the works' for the environmentalists.

But, for one environmentalist at least, Bob Brown, 'the significance of the find left us speechless'. He was one of the party that re-visited the cave with its modern discoverer Kevin Kiernan in 1979. Brown had been searching for a reputed convict skeleton on Goodwin's Peak with the aim of putting recent history to the service of the wilderness cause. A lone skeleton would have been a powerful symbol of the awesome majesty of the south-west, and good publicity for the campaign. But, instead, he happened upon far more ancient bones, those of Ice Age marsupials which Aboriginal people had hunted for food and cooked in a cave. These bones told not of an anguished death in an isolated wilderness, but of hearth and home in a populated landscape. It was not quite the history he had been looking for, but he embraced its message with awe.

This is a shortened version of a paper given in November 1990 at the annual symposium of the Australian Academy of Humanities.

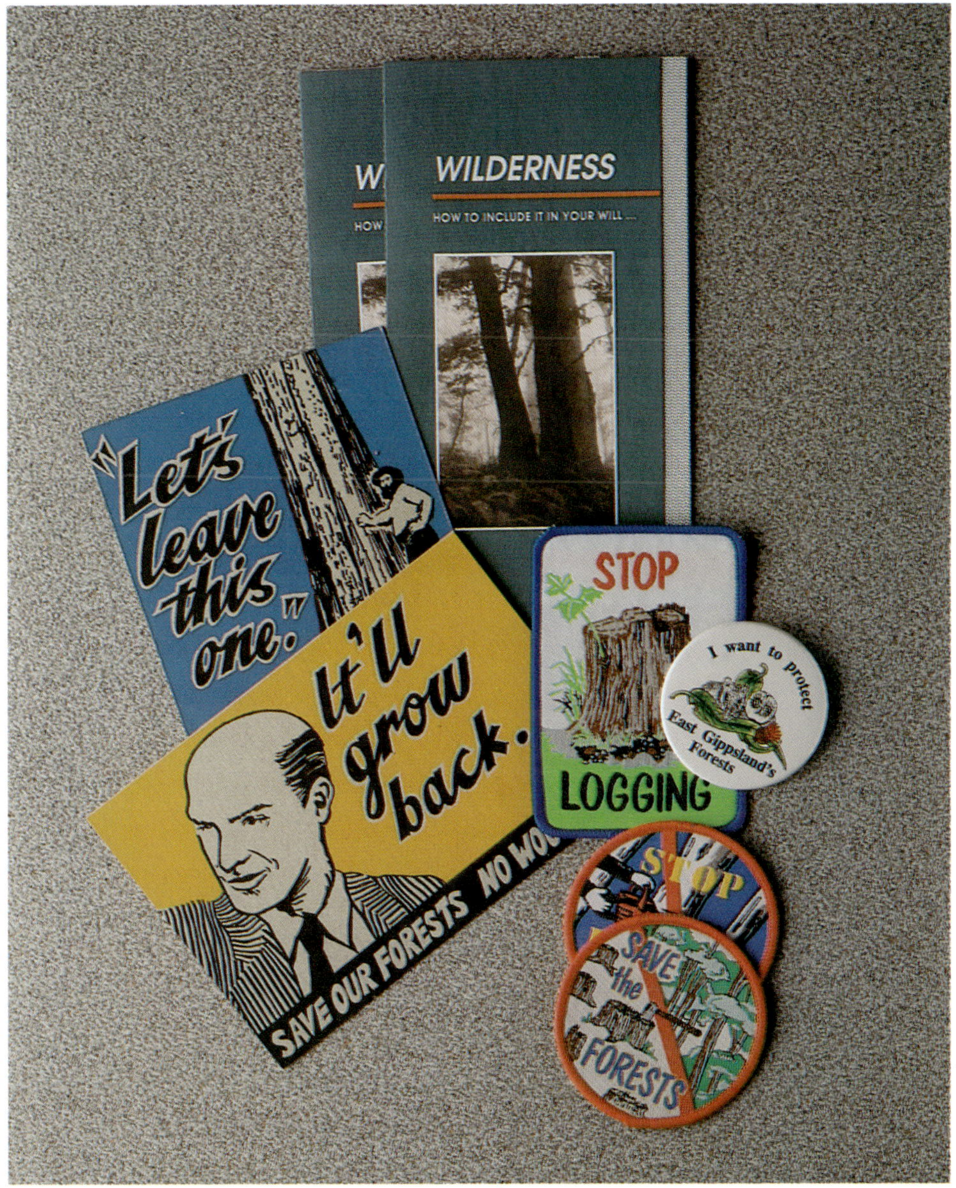

Collection of Anti-logging Ephemera
Photograph: Courtesy of Powerhouse Museum

TERRIBLE PROSPECTS — Geoffrey Batchen

Imagine this. It's the third of March, 1983, two days before a crucial Federal election. At stake is not only the government of Australia, but also the political careers of both the Prime Minister Malcolm Fraser and the new leader of the Labor Party, Bob Hawke. Also at stake is the Franklin River in South-West Tasmania, which is due to be dammed for the production of hydro-electric power by that State's Liberal government. There have been a lot of protests, arrests, and legal challenges, and also a lot of talk about States' rights, unemployment, and the like. The election hangs in the balance. You open your morning newspaper and are confronted by a spectacular full-page colour advertisement. The ad features a reproduction of a landscape photograph, *Rock Island Bend, Franklin River, S.W. Tasmania*, by the Tasmanian nature photographer Peter Dombrovskis, together with a column of text about the dam issue. Across the top of the page runs a provocative question: 'Could you vote for a party that will destroy this?'

It's worth looking more closely at the way this advertisement works. The photograph shows a bend in the Franklin River, with water rushing headlong between two rocky outcrops and then flowing on beyond the bottom of the picture. The surrounding land is rugged, its steep sides glistening with water spray and occasional bands of mosses and ferns. The peaks of the rock outcrops bear a profusion of trees and other foliage. We do not see a lot of this foliage or indeed of the river itself, and scale is difficult to determine. This is partly because the whole scene is swathed in mist and partly because the photographer has chosen to concentrate on just this detail to give us a sense of the whole scene. The image is closely cropped, focusing our attention on the rocky outcrop that stands between the banks on either side of the picture. Our eye goes to this 'island' and is swept beyond it, up the course of the river, and into the misty sky behind. Unable to see clearly, we are forced to imagine what might be there, what scenic wonders are in store, just around the corner.

In every way this is a photograph as much about the unseen as the seen. Notice how we are made to look from a slightly elevated position, as if hovering somewhere above the surface of the earth. We have the privilege of seeing without really being there. There are no signs of our intrusion into this world, no signs of our pathway through its undergrowth to reach this point. It is as if no human being has ever been here before, as if no human being is here now. For our view is the omnipotent view of a God.

There is much else about this image that smacks of the sacred. The cathedral-like architecture of the picture's composition allows the Franklin's vertical rock formations to guide our eyes skyward. The photograph's slow exposure time enhances this reverential quality, giving the flowing water a soft, milky, other-worldly appearance. A sense of silence and timelessness pervades, as if we are seeing a still primeval land, lost in the mists of time. It is a dreamscape, a vision too lovely to be true. Surrounded by a black border, its blue-green colour scheme gleams up at us like a jewel. It is as though an illuminated manuscript has been found hidden within the instrumental, secular cacophony of a daily newspaper. It is a hymn to nature. It is a prayer for life. It is a blessing disguised as a landscape.

FROM THE SUBLIME...

This kind of landscape image was much more common in the nineteenth century than it is today. Back then, the sublime, a theory concerned primarily with the representation of God, was an important inspiration for many artists. An inspiration, but also a frustration, a 'pleasure mixed with pain', since God is unrepresentable. Faced with an impossible task, artists interested in the sublime tended to produce images of a mighty nature towering in an ungraspable vastness over a few tiny awe-inspired human onlookers.

Australian artists found our local rainforest areas offered potential as a backdrop to this image and, from the 1820s onwards, pictures were drawn and painted of the 'impenetrable' tropical vegetation found in places such as Illawarra, the Dandenong Ranges, Wollongong, Leura, Mount Tamborine, and Cairns. Once again these were images as much about absence as presence, in this case the lamentable absence of a desired civilisation.

Rainforest was seen as a wilderness, and therefore as a place of solitude and savageness, an unkempt landscape indelibly marred by its *lack* of cultivation. As one settler confidently predicted in 1821, the rainforest would soon be changed from 'a mournful and desolate wilderness, into the cheerful village, the busy town, the crowded city'. This view of wilderness was reinforced by the biblical story, with its reference to Moses leading his people out of the wilderness and to Christ going into the wilderness for forty days and nights to be tested by the devil. No wonder, then, that wilderness was used as a pejorative term by some Australian commentators. Until 1983, that is.

The advertisement appearing in 1983 was sponsored by the National South-West Coalition, including the Tasmanian Wilderness Society (TWS) and the Australian Conservation Foundation. The TWS had adapted this title in 1976, borrowing its key word from a well-established American counterpart. It borrowed also an American concept of what wilderness is or, at least, of what it ought to be. Bob Brown, who authorised this advertisement, described wilderness in 1982 as 'a region of original Earth where one stands with the senses entirely steeped in Nature and free of the distractions of modern technology'.

The Dombrovskis photograph reproduces an aesthetic version of the same utopian vision. Here, nature has been orchestrated to resemble a twentieth-century concept of wilderness, in which

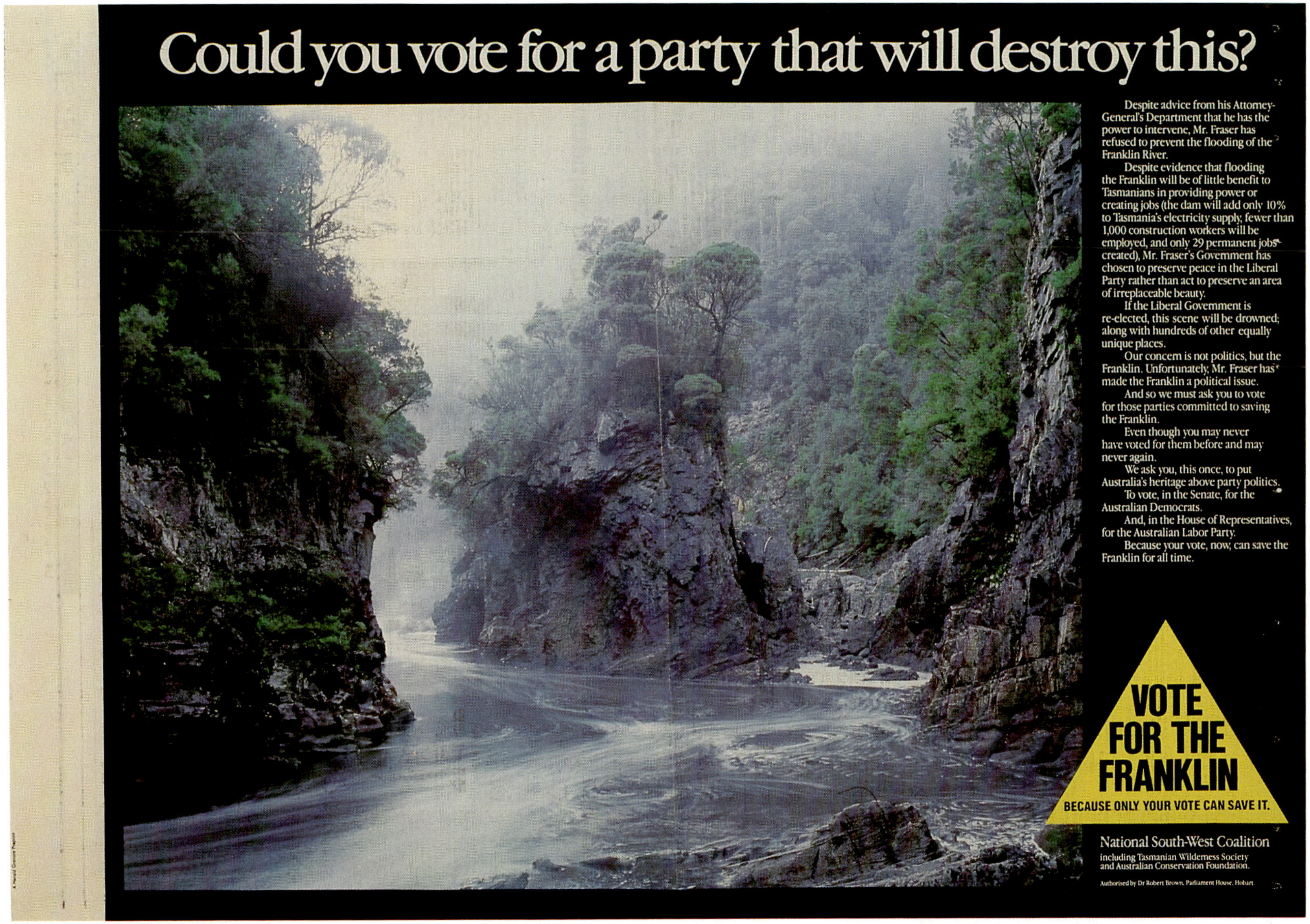

***Vote for the Franklin*, 1983 By Peter Dombrovskis Full page advertisement which appeared in all major Australian newspapers on 3 March 1983**
Copyright: The Wilderness Society/Australian Conservation Foundation Exhibit no. 45

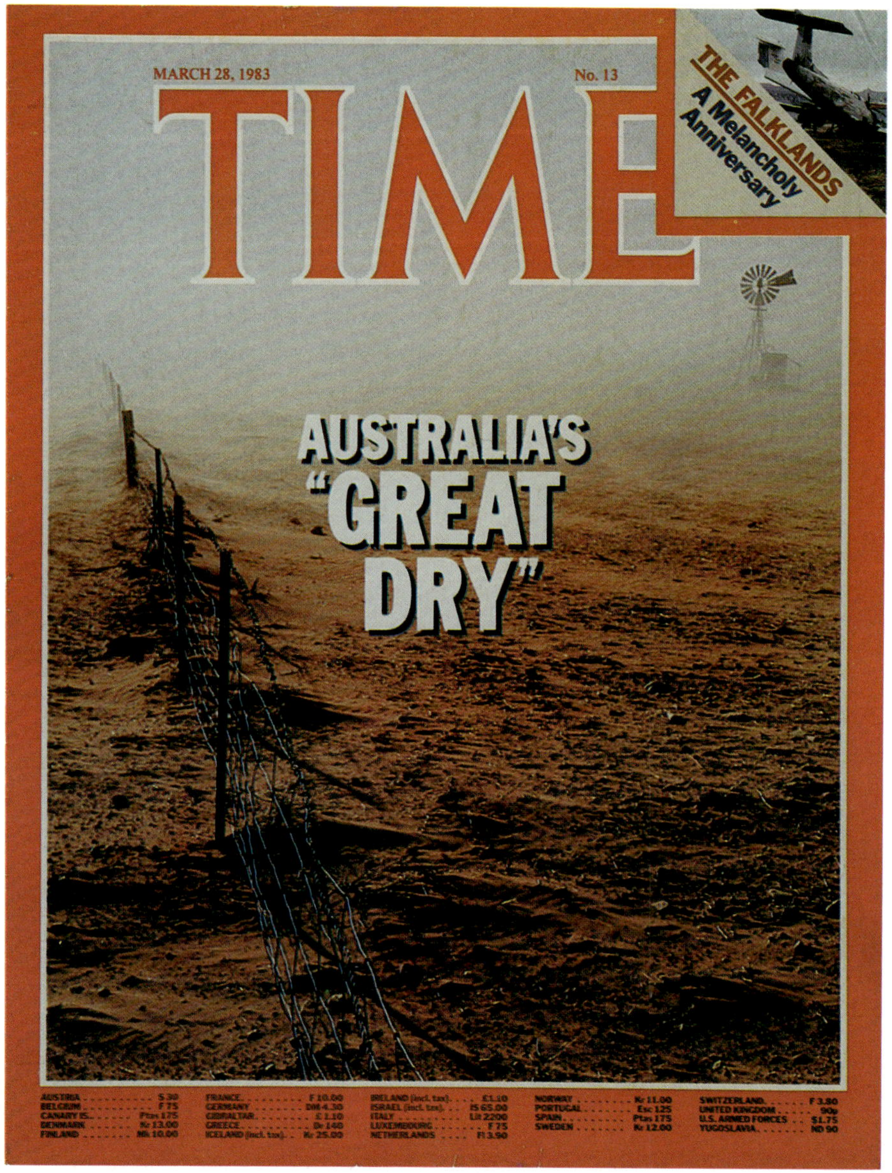

***Time Magazine* front cover, 28 March 1983** **The drought-stricken countryside contrasted dramatically with the Tasmanian wilderness** **Copyright: Time International**

the isolated and the unkempt is celebrated rather than condemned. Wilderness is conceived first and foremost as a sensory experience, as the perception of landscape rather than land.

It is interesting, in this regard, that the word 'wilderness' can also be used to refer to an ornamental garden or park, one laid out as a fantastic maze or labyrinth. Thus, the word itself is haunted by an artful complicity of nature and culture.

So too is the Dombrovskis photograph. This is an image made all the more powerful by its familiarity, having a lush style reminiscent of a thousand *National Geographic* covers already sitting on Australian bookshelves. Full of sentiment and its own moral purity, *Rock Island Bend* comes close to being a cliché of this particular genre of nature photography, an expertly constructed piece of conservation kitsch. Certainly it promises the possibility of a totally pristine environment, untouched by human hand, a promise paradoxically denied by the very existence of this photograph.

However, the promise is more obviously threatened in this advertisement by the yellow triangle in the lower right-hand corner of the image, a loud traffic signal of a logo which recalls the hustle and bustle of the 'crowded city' that this landscape might yet become. Such a threat was all the more poignant in March 1983. The fecundity of this rainforest was in stark contrast to the terrible fires of Ash Wednesday, featured on the 1 March cover of *The Bulletin*, and the drought conditions afflicting the rest of the country, described as 'Australia's "Great Dry" ' in the striking cover for *Time's* 28 March issue. By comparison, the beauty of some verdant uncultivated wilderness was emphasised.

Like its nineteenth-century predecessors, the Save the Franklin advertisement works in terms of the sublime, as a pleasure mixed with pain. The pleasure is provided by the beauty of the Dombrovskis photograph, and all that it signifies. But it is a pleasure sharpened by its potential demise, a terrible prospect stabbed into our consciousness by the banner headline unfurled above. Life and death is laid before us in the form of a simple morality play. And we get to decide the fate of this scene, 'because only your vote can save it'. Who could resist such a challenge? Could we really vote for a party that would destroy this? Well, of course, many of us couldn't.

The next day the government was changed and the proposed dam quickly stopped by federal intervention. So this was one of the rare photographs that has made an almost demonstrable political impact on its viewers. More than that, this advertisement was the climax of a vigorous campaign by conservationists, which continues to inform the way Australians think about their landscape. It not only saved the Franklin; it transformed our notion of 'wilderness', firmly putting this new concept both on the political map and into the popular lexicon of Australian life.

LIST OF EXHIBITS

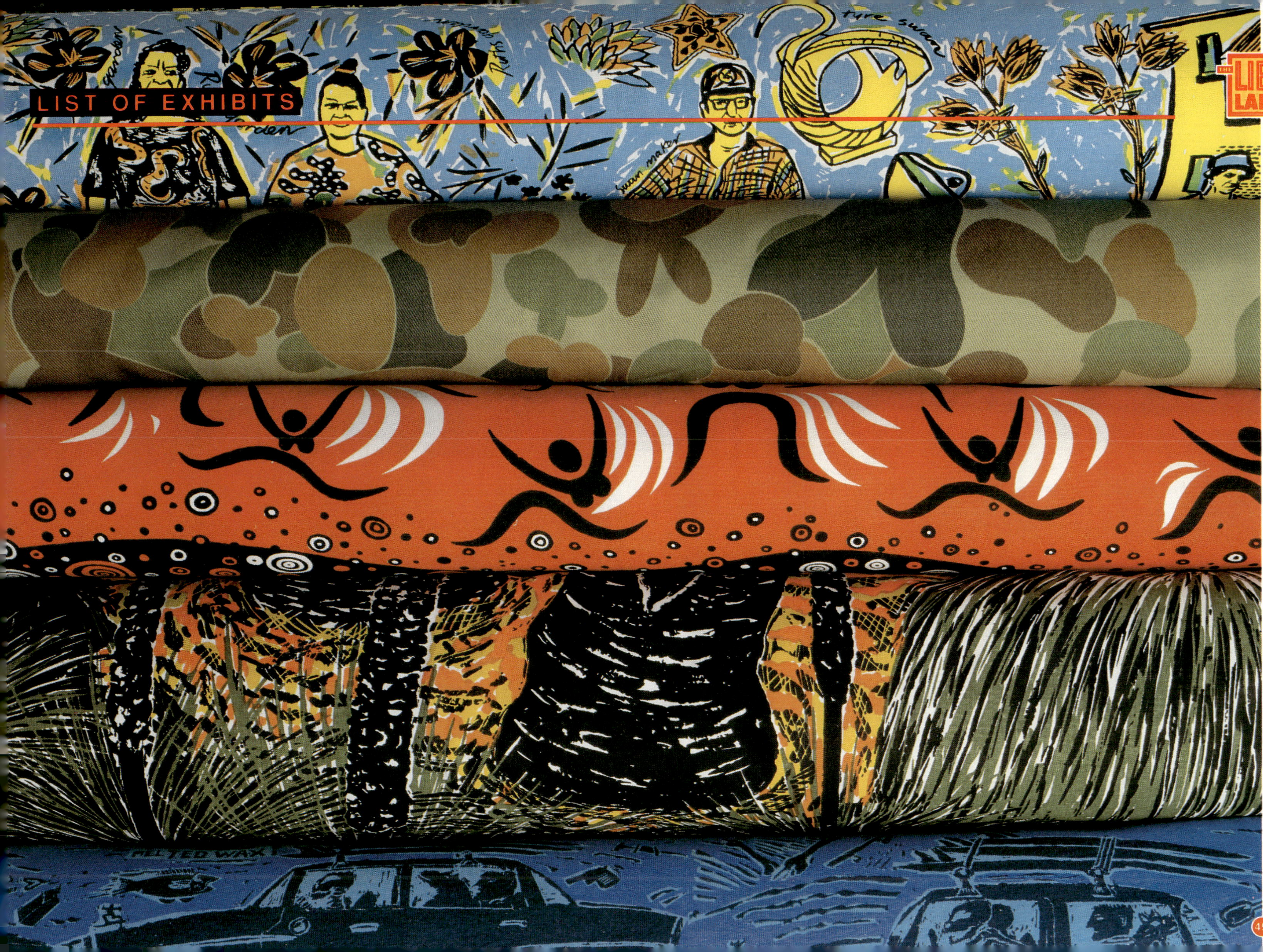

LIST OF EXHIBITS

ICONS

1 *Big Aussie Football Ghost Inspects a Ute-load of Dog-trumpets,* 1990
Surfboard
Artist: Reg Mombassa
Collection: Phil Dwyer

2 *Malibu, Wherever You Are It's a Tropical State,* 1990
Billboard
Designer: Clemenger
Producer: Swifton Moore
Collection: National Centre for Australian Studies

2a Painted Car Door, 1989
Acrylic on BMW car door panel, Papunya, Northern Territory
Artist: Michael Jagamara Nelson
Collection: Courtesy of the Powerhouse Museum
Gift of BMW Australia Ltd 1991

PAINTINGS

3 *The Beach,* 1918
Oil on composition board, 20 x 26
Artist: Elioth Gruner
Collection: University of Sydney Union Art Collection

4 *New England,* 1921
Oil on canvas, 34.3 x 44.5
Artist: Elioth Gruner
Collection: Art Gallery of New South Wales

5 *Fern Gully with Aboriginal Family,* 1863
Oil on canvas, 46.3 x 64.5
Artist: Thomas Clark
Collection: Art Gallery of New South Wales

6* *South Through the Ranges, Heavitree Gap*
Watercolour on paper, 35 x 53
Artist: Albert Namatjira
Collection: Private Collection

6* *November Morning, Simpson's Gap*
Watercolour on paper
Artist: Albert Namatjira
Collection: Private Collection
* To be alternated

TEXTILES

7 Man's Tea-towel Suit, 1989
Designer: Mimmo Cozzolino
Collection: Mimmo Cozzolino

8 *Urban Koori Views,* 1990
Screen-printed cotton
Designer: Alice Hinton-Bateup, Garry Jones & Karen Vance, Garage Graphix
Collection: Courtesy of the Powerhouse Museum

9 *Black Boys,* 1987
Screen-printed cotton
Designer: Paul Worstead, House of Feral, Australia
Collection: Courtesy of the Powerhouse Museum

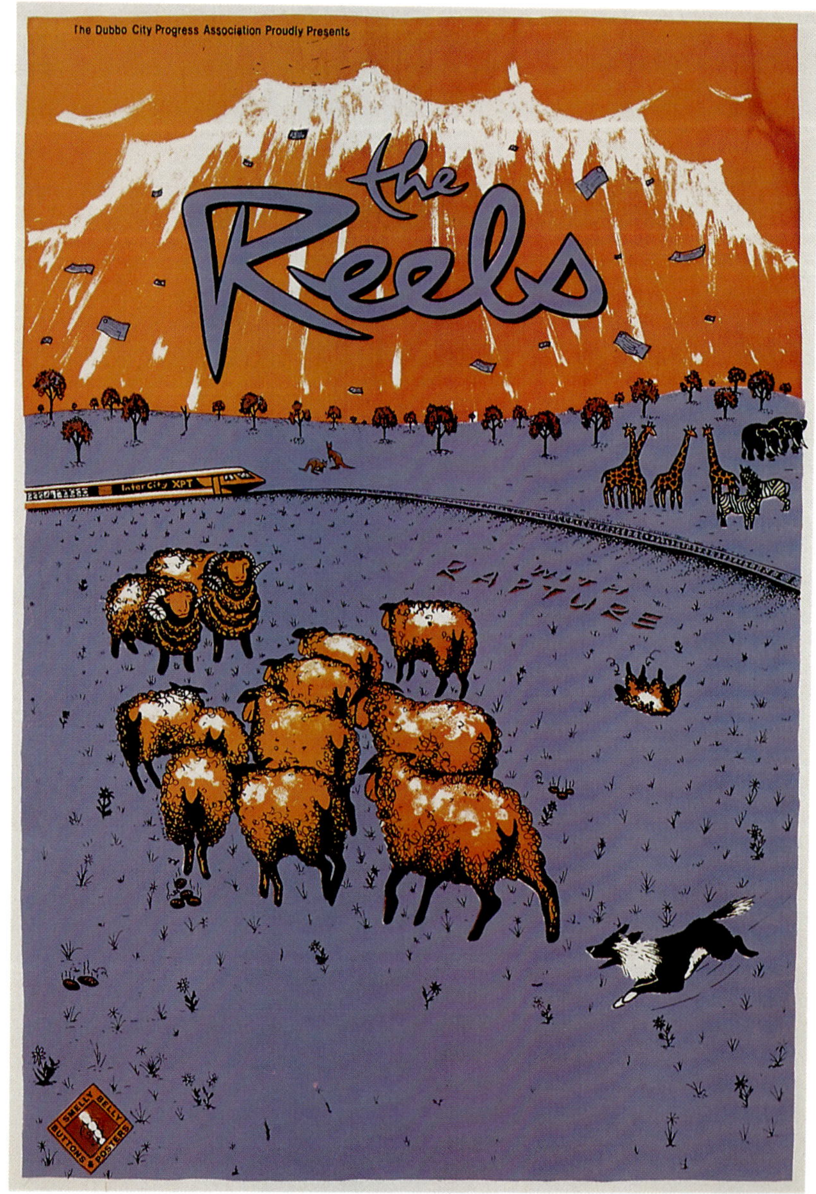

The Reels **By Libby Blainey, Smelly Belly Posters & Buttons Exhibit no. 38**

LIST OF EXHIBITS

10 ***Houses and Gardens,*** 1990
Screen-printed cotton
Designer: Jan McKay
Collection: Courtesy of the Powerhouse Museum

11 ***Camouflage,*** 1987
Computer-designed cotton/polyester
Designer: M R L Optics Research Group for Army combat uniforms
Collection: Courtesy of the Powerhouse Museum

12 ***Credibility Strip,*** 1988
Screen-printed cotton
Designer: Gerry Wedd, Mambo
Collection: Courtesy of the Powerhouse Museum

POSTERS

COAST

13 ***Vast,*** 1988
Offset, 100 x 70
Photographer: Branco Gaica
Producer: Sydney Dance Company
Collection: Courtesy of the Powerhouse Museum

14 ***The Gold Coast Queensland,*** 1990
Offset, 96 x 60
Designer: Scanlon
Producer: Australian Airlines
Collection: National Centre for Australian Studies

15 ***Behind Enemy Lines,*** 1981
Silk-screen, 56 x 76
Artist: Jan Fieldsend
Collection: National Centre for Australian Studies

16 ***The Best Thing next to a Suntan,*** 1983
Offset
Producer: Speedo Australia
Collection: Courtesy of the Powerhouse Museum

17 ***No Nukes in the Pacific,*** 1984
Silk-screen, 65 x 91
Artist: Pam Debenham
Collection: National Centre for Australian Studies

18 ***You Are on Aboriginal Land,*** 1987
Silk-screen, 76 x 55
Designer: Marie McMahon, (4th edition)
Printer: Redback Graphix
Collection: Courtesy of the Powerhouse Museum

19 ***The Spring Flower Show,*** 3 September 1991
Newsprint: *The Sydney Morning Herald*, 60 x 42
Producer: David Jones
Collection: Ann Stephen

20 ***Help,*** 9 July 1991
Newsprint: *The Age*, 60 x 42
Producer: Esso Australia Ltd
Collection: National Centre for Australian Studies

21 ***Clean Air,*** c.1985
Offset, 50 x 75
Producer: NSW State Pollution Control Commission
Collection: Jai Paterson

22 ***The New Deal,*** 1985
Offset, 68 x 97
Artist: Richard Allan, Mambo
Collection: Courtesy of the Powerhouse Museum

SUBURBS

23 ***Trespassers Keep Out,*** 1982
Silk-screen, 51 x 76
Artist: Avril Quail
Collection: University Art Workshop

24 ***Win a House,*** 6 August 1991
Newsprint: *The Daily Telegraph Mirror*, 60 x 40
Publisher: Nationwide News, Surry Hills
Collection: Ann Stephen

25 ***Victoria Street, Brunswick,*** 1986
Silk-screen, 51 x 76
Artist: David Morrow, Redletter Press (now Red Planet)
Collection: National Centre for Australian Studies

26 ***Ulladulla,*** 1990
Offset, 62 x 43
Designer: Allan Droscher, My Town Cartoon Maps
Printer: Weston & Co. Print, Kiama
Collection: National Centre for Australian Studies

27 ***More a Part of the Landscape than a Pair of Trousers,*** 1988
Offset, 63 x 46
Artist: Chris O'Doherty (alias Reg Mombassa), Mambo
Collection: Courtesy of the Powerhouse Museum

28 ***The Park Estate, Rockdale,*** 1924
Lithograph, 85 x 65
Designer: F. Cunninghame & Co. Ltd, Sydney
Collection: Courtesy of the Powerhouse Museum

29 ***Mount Druitt — A Bit of the Country and a Bit of the City,*** 1987
Calendar page, silk-screen, 51 x 38
Artist: Garage Graphix
Collection: National Centre for Australian Studies

30 ***New Choices in Housing,*** 1990
Offset, 82 x 44
Producer: Commonwealth Department of Health, Housing & Community Services
Collection: National Centre for Australian Studies

31 ***A Car, 2 Kids, a Dog and... a VDT,*** 1984
Silk-screen, 76 x 67
Artist: J. Clutterbuck
Producer: Australian Insurance Employee's Union
Collection: Courtesy of the Powerhouse Museum

LIST OF EXHIBITS

32 *We Wish All these Dreams Could Come True,* 1990
Silk-screen, 75 x 50
Artists: Jenny Pitty and Penny Richardson,
Producer: Garage Graphix, D. Cazalet, Villawood Community Artswork and women residents of Villawood
Collection: National Centre for Australian Studies

33 *Windy Gully,* 1989
Silk-screen, 75 x 102
Artist: Gregor Cullen
Producer: Theatre South, Wollongong
Collection: National Centre for Australian Studies

PASTORAL

34 *Heritage Council of New South Wales,* 1982
Offset, 36.5 x 48.6
Producer: Heritage Council of New South Wales
Collection: Jai Paterson

35 *Sunlover Holidays Queensland*
Offset, 64 x 47
Producer: Unknown
Collection: National Centre for Australian Studies

36 *S.O.S.*
Silk-screen, 65 x 72
Designer: Julie Lynch,
Producer: Salamanca Theatre Company, Tasmania
Collection: Tasmaniana Collection, State Library of Tasmania

37a/b *NSW is Larger than Life*
Offset, 40 x 100
Producer: NSW Tourism Commission
Collection: National Centre for Australian Studies

38 *The Reels*
Silk-screen, 76 x 52
Artist: Libby Blainey, Smelly Belly Posters & Buttons
Collection: Libby Blainey

39 *Another Story,* 1988
Silk-screen, 80 x 60
Artist: Sally Morgan
Producer: Australian Bicentennial Authority
Collection: Courtesy of the Powerhouse Museum

40 *A Land of Sweeping Plains,* 1988
Offset, 50 x 70
Artists: Mark Lang, Adrian Davey, Marie Lochman, Wade Hughes, Natfoto, Otto Rogge, Australasian Nature Commission, Lochman Transparencies
Producer: Australian Heritage Commission
Collection: Australian Heritage Commission

41 *Ford Corsair,* 1991
Offset, 92 x 69
Producer: Ford Motor Co. of Australia Ltd
Collection: Ford Motor Co. of Australia Ltd

RAINFOREST

42 *Cascade Beer,* 1991
Offset, 27 x 19
Artist: Otto Schmidinger
Producer: Hertz Walpole Advertising
Collection: Cascade Brewing Company

43 *Forests' Disaster Rally,* 1985
Offset, 50 x 38
Collection: Tasmaniana Collection, State Library of Tasmania

44 *1982 Wilderness,* 1982
Calendar, offset, 44 x 28
Designer: Kalianna Press, Chatswood
Photographer: Henry Gold
Producers: National Parks Association of NSW & The Colong Committee Ltd, a National Wilderness Society
Collection: Sandra McEwen

45 *Vote for the Franklin,* 3 March 1983
Newsprint: *The Sydney Morning Herald,* 42 x 60
Photographer: Peter Dombrovskis
Producer: National South-West Coalition including Tasmanian Wilderness Society and the Australian Conservation Foundation
Collection: Ian Burn

46 *Mossman Gorge,* 1990
Offset, 96 x 60
Designer: Scanlon
Producer: Australian Airlines
Collection: National Centre for Australian Studies

47 *Save the Franklin, Damn the Government,* 1981
Silk-screen, 76 x 51
Artist: Bob Clutterbuck
Producer: Redletter Press (now Red Planet)
Collection: National Centre for Australian Studies

48 *Don't Log Rainforest,* 1981
Silk-screen, 76 x 51
Artist: Angela Gee
Collection: Sydney University Workshop

49 *Australian Wilderness Heritage*
Offset, 60 x 80
Photographer: David Tatnall
Designer: David Neilson
Producer: The Wilderness Society
Collection: Courtesy of the Powerhouse Museum

50 *Tomasetti Goes Green,* 1990
Offset, 84 x 60
Producer: Tomasetti Paper
Printer: Incolour Printing
Collection: Allan Coggin

51 *A Typical Logged Forest,* 1991
Tram Advertisement, 61 x 305.5
Producer: Victorian Timber Industry
Collection: National Centre for Australian Studies

LIST OF EXHIBITS

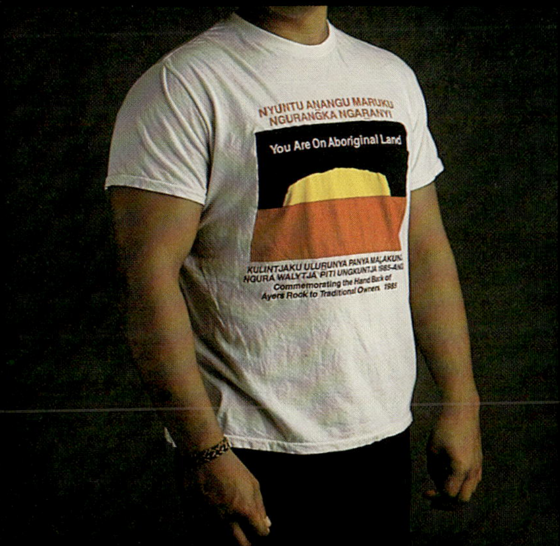

You Are on Aboriginal Land, 1985 By Chips Mackinolty Copyright: Mut it julu Community Exhibit no. 67

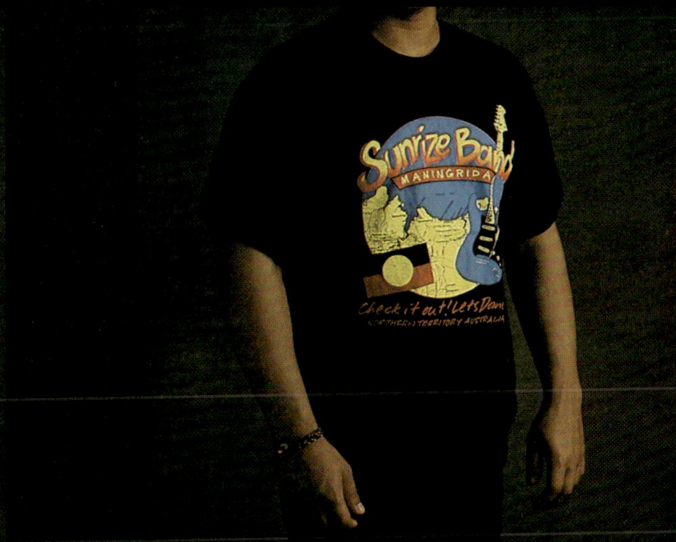

Sunrize Band, 1990 Designed by Todd Williams Exhibit no. 72

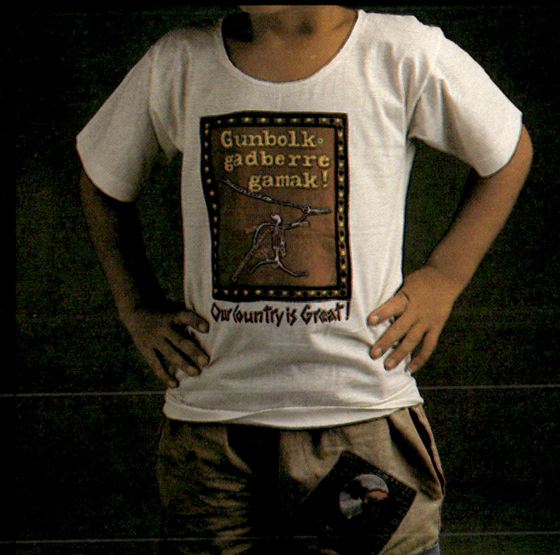

Our Country Is Great, 1987 Designed by Chips Mackinolty Exhibit no. 70

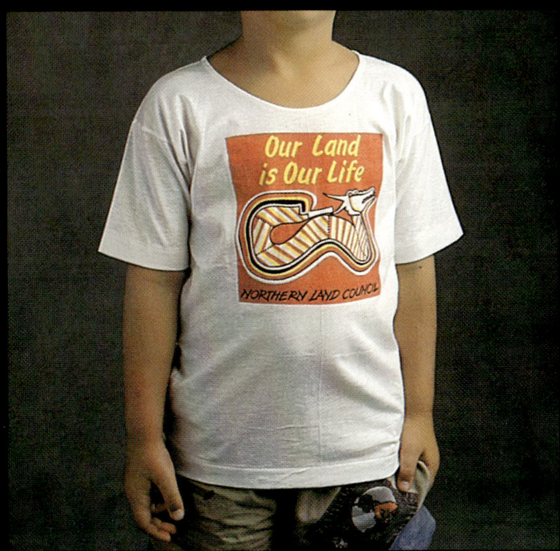

Our Land Is Our Life, 1987 Designed by Curly Bardugupu Jalak Exhibit no. 69

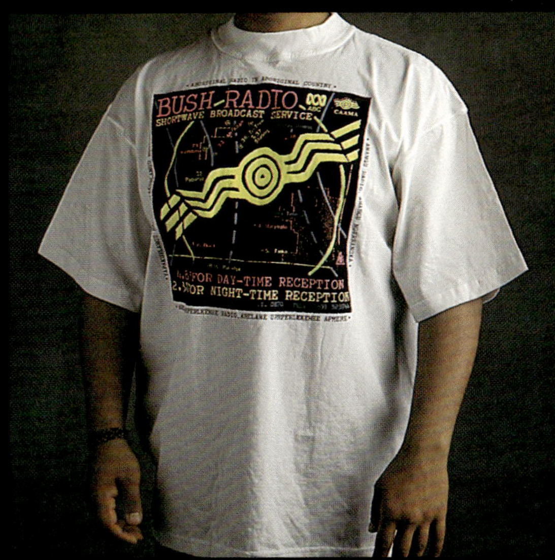

Bush Radio, 1986 Designed by Michael Callaghan, Redback Graphix Exhibit no. 68

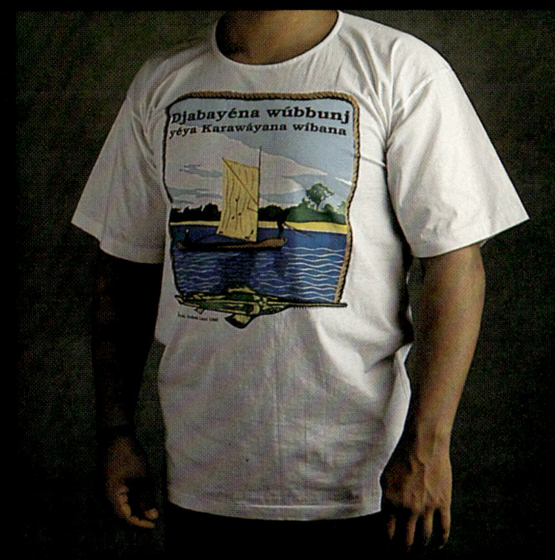

Manayingkarinra, 1990 Designed by Chips Mackinolty Exhibit no. 73

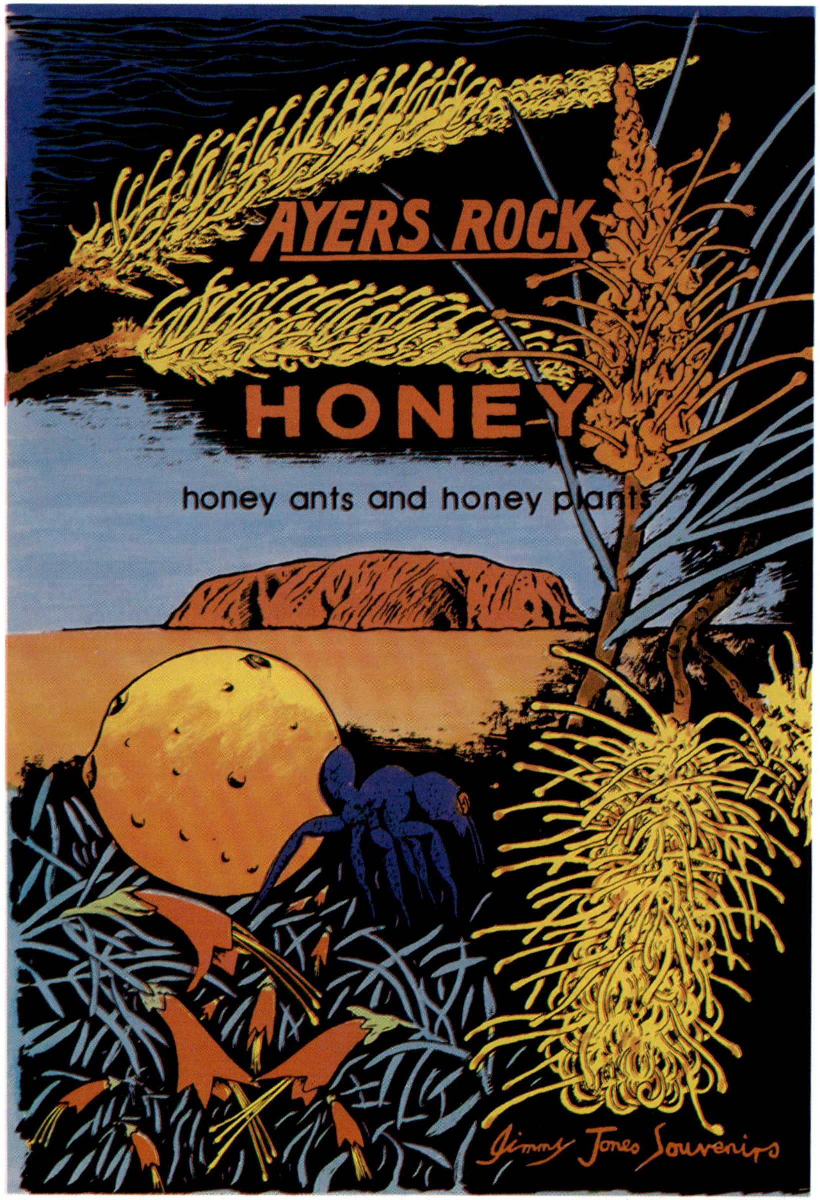

Ayers Rock Honey, 1980 Postcard by Jimmy Jones Souvenirs (Paul Worstead)

OUTBACK

52 *Bush Medicine*, 1991
Offset, 86 x 60
Photographers: Terry Knight 'Passing the Knowledge' & John Brock, Clyde Dunlop, Nick Smith, Glenn Wightman
Producer: Northern Territory Conservation Commission
Collection: National Centre for Australian Studies

53 *Beat the Grog*, 1986
Silk-screen, 63 x 94
Artist: Michael Callaghan, Redback Graphix
Producer: Commissioned by Central Australian Aboriginal Media Association
Collection: Courtesy of the Powerhouse Museum

54 *History Story*, 1990
Offset, 70 x 50
Artist: Alec Minutjukur
Photographer: David Haigh
Text: Alex Minutjukur & Ely White
Producer: Department of Aboriginal Affairs
Collection: Courtesy of the Powerhouse Museum

55 *You Can't Beat the Feeling*, 1991
Offset, 50 x 90
Designer: McCann-Erickson, Sydney
Producer: Coca-Cola South Pacific
Collection: Ann Stephen

56 *Land Rights News*, July 1988
Tabloid, Newsprint, 42 x 29
Artist: Clem Abbott
Producer: Northern Territory Land Councils
Collection: National Centre for Australian Studies

57 *The Pilbara*
Offset, 86 x 56
Producer: Western Australian Tourism Commission
Collection: National Centre of Australian Studies

58 *Mad Max — Beyond Thunderdome*, 1985
Offset, 100 x 65
Artist: Amsel
Producer: Warner Bros.
Collection: National Centre of Australian Studies

59 *There is Nowhere in the World Like Your Own Territory*
Offset, 48 x 69
Producer: Unknown
Collection: National Centre for Australian Studies

60 *Two Laws Together*, 1991
Silk-screen, 51 x 150
Artist: Wenten Rubuntja & Chips Mackinolty
Producer: Aboriginal Areas Protection Authority
Collection: National Centre for Australian Studies

LIST OF EXHIBITS

T-SHIRTS

61 Two Hundred Years of T-shirts, 1980
Designer: Paul Worstead
Collection: National Centre for Australian Studies

62 Ayers Rock Honey, 1980
Designer: Paul Worstead
Collection: National Centre for Australian Studies

63 Desert Rose, 1980
Designer: Paul Worstead
Collection: National Centre for Australian Studies

64a/b Ayers Rock Bread (left & right), 1981
Designer: Paul Worstead
Collection: National Centre for Australian Studies

65 Ayers Rock Fruit Salad, 1981
Designer: Paul Worstead
Collection: National Centre for Australian Studies

66 I Climbed Ayers Rock, 1984
Designer: Paul Worstead
Collection: National Centre for Australian Studies

67 You Are on Aboriginal Land, 1985
Designer: Chips Mackinolty, Copyright Mut it julu Community
Collection: Courtesy of the Powerhouse Museum

68 Bush Radio, 1986
Designer: Michael Callaghan, Redback Graphix
Collection: National Centre for Australian Studies

69 Our Land Is Our Life, 1987
Designer: Curly Bardugupu, Jalak
Collection: National Centre for Australian Studies

70 Our Country Is Great, 1987
Designer: Chips Mackinolty, Jalak
Collection: National Centre for Australian Studies

71 Care for this Country, 1988
Designer: Alison Alder, Redback Graphix
Collection: National Centre for Australian Studies

72 Sunrize Band, 1990
Designer: Todd Williams
Collection: National Centre for Australian Studies

73 Manayingkarinra, 1990
Designer: Chips Mackinolty
Collection: National Centre for Australian Studies

74 Gurindji Freedom Day, 1966-1991, 1991
Designer: Chips Mackinolty, Green Ant
Collection: National Centre for Australian Studies

TEA-TOWELS

75 The Australian Swagman, c.1990
Designer: Souvenirs Australia
Collection: National Centre for Australian Studies

76 Australian Rainforest, c.1990
Designer: Ross
Collection: National Centre for Australian Studies

77 Forest Not Woodchip, c.1990
Designer: Nosey Posters
Collection: National Centre for Australian Studies

78 The Whitsundays
Designer: Australian Souvenirs & Marketing P/L
Collection: National Centre for Australian Studies

79 Ka-poo, 1989
Designer: Eliza Campbell & Judith Lodwick
Collection: National Centre for Australian Studies

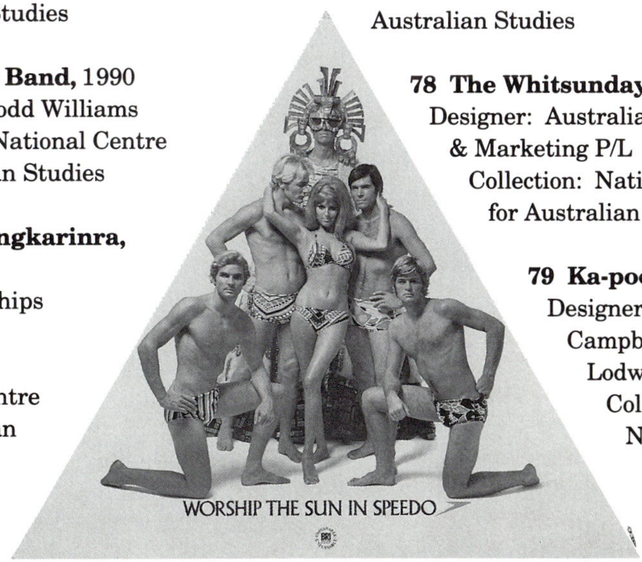

Worship the Sun in Speedo **Point of sale Speedo advertisement**

80 Here's 4 Great Reasons for Food Irradiation, c.1989
Designer: Eliza Campbell & Judith Lodwick
Collection: National Centre for Australian Studies

81a/b Older than the Hills, 1988
Designer: Eliza Campbell & Judith Lodwick
Collection: National Centre for Australian Studies

82 Come Back to Catholicism, 1986
Designer: Eliza Campbell & Judith Lodwick
Collection: National Centre for Australian Studies

SNOWDOMES

COAST

83 Surfing Lakes Entrance
Collection: State Library of Victoria
84 Lakes Entrance Vic
Collection: Wayne Golding
85 Greetings from Batemans Bay
Collection: Wayne Golding
86 Rottnest Island WA
Collection: Wayne Golding
87 Rottnest Island WA
Collection: Wayne Golding
88 Great Barrier Reef Qld
Collection: State Library of Victoria
89 Great Barrier Reef
Collection: Wayne Golding
90 The Sunshine Coast Queensland
Collection: State Library of Victoria
91 Big Banana Coffs Harbour
Collection: State Library of Victoria

LIST OF EXHIBITS

92 Banana Bender Coffs Harbour
Collection: Wayne Golding
93 Tangalooma Your Holiday Island
Collection: Wayne Golding
94 Hervey Bay QLD
Collection: Tracey Naughton
95 Penguin Parade
Collection: Annette Welkamp
96 Greetings from Bondi Beach
Collection: Wayne Golding
97 Narooma
Collection: Wayne Golding
98 Greetings from Surfers Paradise Queensland
Collection: Wayne Golding
99 Gold Coast
Collection: Wayne Golding
100 Mandurah W. A. Blue Manna Country
Collection: Wayne Golding
101 Nelson Bay
Collection: Wayne Golding
102 Port Macquarie NSW
Collection: Wayne Golding
103 Lake Moogerah Heart of the Fassifern
Collection: Wayne Golding
104 Gympie. Q.
Collection: Wayne Golding

SUBURBS

105 Telecom Tower Canberra
Collection: State Library of Victoria
106 Opera House Sydney
Collection: State Library of Victoria
107 Arts Centre Melbourne Victoria
Collection: Annette Shiell
108 Arts Centre Melbourne Victoria
Collection: Wayne Golding
109 Melbourne
Collection: Tracey Naughton
110 Parliament House Canberra
Collection: Annette Welkamp
111 Parliament House Canberra
Collection: Wayne Golding

PASTORAL

112 Australia
Collection: Tracey Naughton
113 Armidale
Collection: Tracey Naughton
114 The Big Merino Goulburn NSW
Collection: Annette Welkamp
115 The Big Merino Goulburn NSW
Collection: Wayne Golding
116 Dog on the Tuckerbox Gundagai NSW
Collection: Annette Welkamp
117 Suncoast Dairy Sunshine Coast, Qld
Collection: Wayne Golding
118 Mittagong, NSW
Collection: Wayne Golding

OUTBACK

119 Broken Hill the Silver City
Collection: Wayne Golding
120 Ayers Rock
Collection: Jacky Talbot
121 Northern Territory Australia
Collection: State Library of Victoria
122 Northern Territory Australia
Collection: Wayne Golding
123 Darwin N.T.
Collection: Wayne Golding

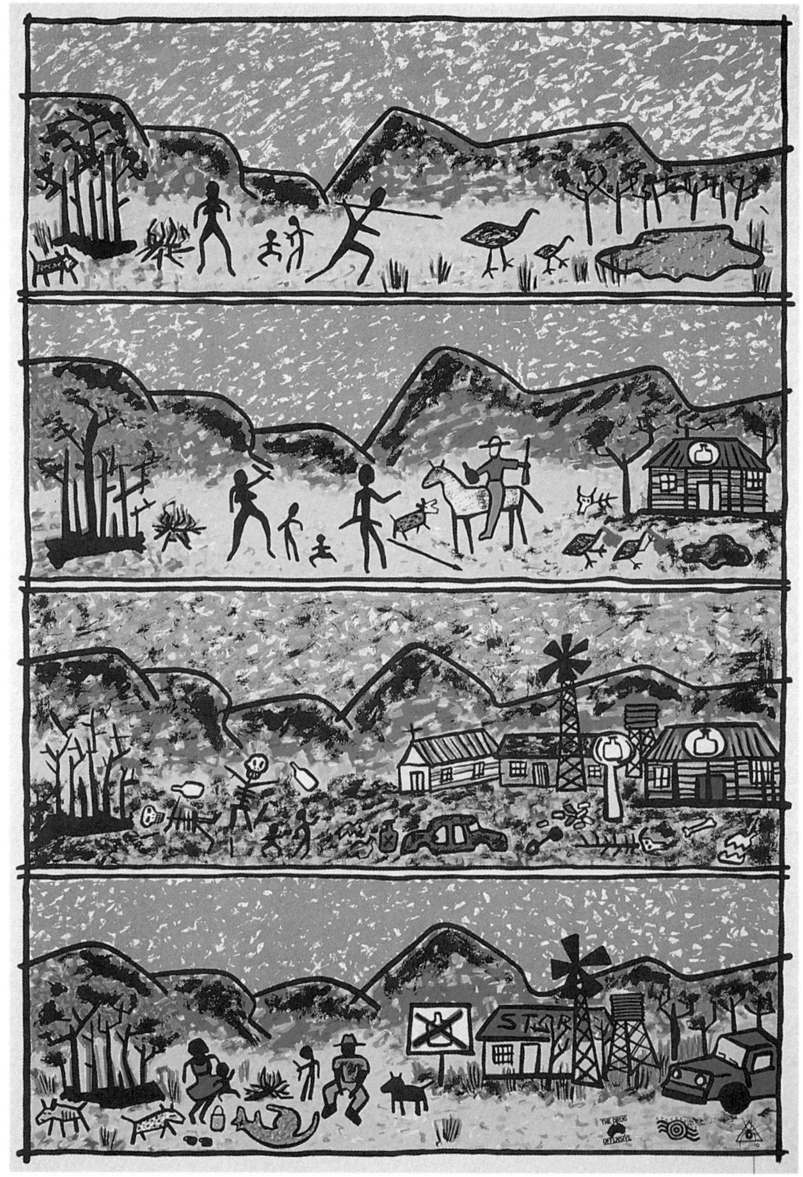

***Beat the Grog*, 1986 Designed by Michael Callaghan Exhibit no. 53**

LIST OF EXHIBITS

124 The Grampians VIC
Collection: State Library of Victoria
125 Kangaroo
Collection: State Library of Victoria
126 Kangaroo
Collection: Annette Welkamp
127 Australian Koalas
Collection: Wayne Golding
128 Australian Kangaroos
Collection: Wayne Golding

OTHER

129 Western Australia Black Swan
Collection: Wayne Golding
130 Batehaven Birdland
Collection: Wayne Golding
131 Perth
Collection: Wayne Golding
132 P. S. Gem, Swan Hill
Collection: Wayne Golding
133 Cooper Pedy South Australia
Collection: Wayne Golding
134 Tasmanian Tigers
Collection: Tracey Naughton
135 Tasmania
Collection: Wayne Golding
136 Tasmania
Collection: Wayne Golding
137 Mount Buffalo Chalet Vic
Collection: State Library of Victoria
138 Jenolan Caves NSW
Collection: Wayne Golding
139 The World's Largest Trout, the Snowy Mountains, NSW
Collection: Wayne Golding
140 Mt. Isa, Copper City
Collection: Wayne Golding
141 Katoomba 3 Sisters
Collection: Wayne Golding

Constructed of plastic, these snowdomes, usually commissioned by Australian souvenir wholesalers, were predominantly manufactured in Hong Kong, c. 1960–91.

In addition to the individually catalogued items the exhibition includes:

VIDEO PROGRAM

Divided into the five landscape sites, this includes a selection of tourist promotion and product advertisements, rock video clips, and an Aboriginal children's program. The material selected was shown on Australian television at some time between 1980 and 1991.

COLLECTION OF BUMPER STICKERS

Collected during 1991, these stickers are a representative sample of the promotional material produced by tourist, youth, industrial, political and social action groups.

COLLECTION OF POSTCARDS

This features tourist postcards produced c. 1960–91 and also includes artist-produced postcard sets by:

- Jon Paton (Wilderness Society, 1991)
- Robert Pearce (A Man About the House, 1982)
- Ruth Waller (House About Wollongong, 1985)
- Paul Worstead (Jimmy Jones Souvenirs) for the Mupitjulu Community

COLLECTION OF FIRST DAY COVERS

Featuring landscape and Aboriginal imagery, c. 1970–90

COLLECTION OF EPHEMERA

This includes displays of:

- felt pennants depicting the attractions of particular tourist destinations
- mass-produced souvenirs including spoons, coasters, bookmarks, badges, salt and pepper shakers, money boxes
- coastal items including advertising material and swimwear for Speedo and Ken Done
- food packaging from companies which manufacture a range of products including jam, oil, fruit juice and iceblocks
- copies of *Landrights News*
- greeting cards featuring suburban imagery, produced by a number of local action groups, c. 1985–90
- brochures, catalogues and advertising material produced for mass distribution by the tourism and manufacturing industries, the housing industry and a variety of political groups, c. 1990–91

NOTES ON CONTRIBUTORS

Geoffrey Batchen is an Australian cultural critic who is currently teaching in the Visual Arts Department at the University of California, San Diego.

Marjorie Bilbil is a Maaridiamu woman, born in the 1940s at Knuckey's Lagoon south of Darwin. She lives with her people at Belyuen, across the Darwin Harbour, and is an experienced adult educator.

Ian Burn is an artist and writer who lives in Sydney. He is the author of *National Life and Landscapes — Australian Painting 1900-1940* which was published in 1991.

Mimmo Cozzolino is a Melbourne-based graphic designer and author. He is also an amateur collector and recorder of Australian ephemera. His book *Symbols of Australia* was first published in 1980.

Ross Gibson is a Sydney-based writer and critic who also teaches Screen Studies at the University of Technology, Sydney.

Tom Griffiths is a lecturer in Public History at Monash University and also worked as a historian at the Department of Conservation and Environment, Victoria. His book *Beechworth: An Australian Country Town and its Past* was published in 1987.

Anthony Hayes completed his Honours Degree in History at the University of Queensland in 1988, and is now Sports Marketing Executive for Qantas' Multi-National Marketing Department.

Sherry McLaughlin is an adult educator from Geraldton, Western Australia.

Richard Peck is the Curator of Printed Technology, Philately and Postal Technology at the Powerhouse Museum. He was a member of the Australia Post Advisory Committee between 1985-1990.

Kathleen Phillips works at the Powerhouse Museum as a video producer. She has had an active interest in popular music for more than 20 years.

Charles Pickett is a Curator of Social History at the Powerhouse Museum. His book *Refreshing! Art Off the Pub Wall* was published in 1990.

Peter Robinson both coordinates and lectures in the Graduate Diploma in Australian Tourism at Monash University. He is also researching the history of Victorian tourism.

Gawen Rudder has worked in Australian marketing, public relations and advertising for 25 years, the past 15 with Chiat/Day/Mojo.

Terry Sams lives and works in Peppimenarti, Northern Territory, as an adult educator. He is involved in music and is presently forming a community band.

Annette Shiell is the Exhibition Coordinator at the National Centre for Australian Studies and Co-curator of **The Lie of the Land** exhibition.

Katie Spearritt previously worked for the National Centre for Australian Studies, before joining the Corporate Public Affairs Department of BHP.

Peter Spearritt is the Director of the National Centre for Australian Studies. He is the author of *The Sydney Harbour Bridge: A Life* (1982) and co-editor of *Packaging the Past? Public Histories* (1991).

Ann Stephen is a Curator of Social History at the Powerhouse Museum and Co-curator of the **Lie of the Land** exhibition. She is co-author of *Badges of Labour, Banners of Pride* (1984) and *The Necessity of Australian Art* (1988).

John Tiparui from Nguiu, Tiwi Islands, is a community health worker specialising in alcohol awareness programs.

Anne-Marie van de Ven is a Curator of Australian Decorative Arts & Design at the Powerhouse Museum.

Kimberley Webber is a Curator of Social History at the Powerhouse Museum and a long-time resident of suburbia.

Galarruwy Yunupingu, AM, comes from the Gumatji clan in East Arnhem Land. He joined the Northern Land Council in 1975 and is currently serving his fourth term as Chairman.